AF794814

Metal-Organic Frameworks With Heterogeneous Structures

Scrivener Publishing
100 Cummings Center, Suite 541J
Beverly, MA 01915-6106

Publishers at Scrivener
Martin Scrivener (martin@scrivenerpublishing.com)
Phillip Carmical (pcarmical@scrivenerpublishing.com)

Metal-Organic Frameworks With Heterogeneous Structures

Ali Morsali and Kayhaneh Berijani

Department of Chemistry, Tarbiat Modares University, Tehran, Islamic Republic of Iran

Scrivener Publishing

WILEY

This edition first published 2021 by John Wiley & Sons, Inc., 111 River Street, Hoboken, NJ 07030, USA and Scrivener Publishing LLC, 100 Cummings Center, Suite 541J, Beverly, MA 01915, USA
© 2021 Scrivener Publishing LLC
For more information about Scrivener publications please visit www.scrivenerpublishing.com.

All rights reserved. No part of this publication may be reproduced, stored in a retrieval system, or transmitted, in any form or by any means, electronic, mechanical, photocopying, recording, or otherwise, except as permitted by law. Advice on how to obtain permission to reuse material from this title is available at http://www.wiley.com/go/permissions.

Wiley Global Headquarters
111 River Street, Hoboken, NJ 07030, USA

For details of our global editorial offices, customer services, and more information about Wiley products visit us at www.wiley.com.

Limit of Liability/Disclaimer of Warranty
While the publisher and authors have used their best efforts in preparing this work, they make no representations or warranties with respect to the accuracy or completeness of the contents of this work and specifically disclaim all warranties, including without limitation any implied warranties of merchantability or fitness for a particular purpose. No warranty may be created or extended by sales representatives, written sales materials, or promotional statements for this work. The fact that an organization, website, or product is referred to in this work as a citation and/or potential source of further information does not mean that the publisher and authors endorse the information or services the organization, website, or product may provide or recommendations it may make. This work is sold with the understanding that the publisher is not engaged in rendering professional services. The advice and strategies contained herein may not be suitable for your situation. You should consult with a specialist where appropriate. Neither the publisher nor authors shall be liable for any loss of profit or any other commercial damages, including but not limited to special, incidental, consequential, or other damages. Further, readers should be aware that websites listed in this work may have changed or disappeared between when this work was written and when it is read.

Library of Congress Cataloging-in-Publication Data

ISBN 978-1-119-79204-8

Cover image: Pixabay.com
Cover design by Russell Richardson

Set in size of 12pt and Minion Pro by Manila Typesetting Company, Makati, Philippines

10 9 8 7 6 5 4 3 2 1

FSC www.fsc.org
MIX
Paper from responsible sources
FSC® C013604

Contents

List of Illustrations **vii**

List of Tables **xv**

List of Schemes **xvii**

Preface **xix**

Abbreviations **xxi**

1 Introduction: A Brief Introduction About Metal-Organic Frameworks **1**
- 1.1 Metal-Organic Frameworks 1
- 1.2 Conclusion 8
- References 8

2 Metal-Organic Frameworks Complexity **13**
- 2.1 Perspectives on Complexity in MOFs 13
- 2.2 Conclusion 21
- References 22

3 Complexity Based on Ligand—Part 1 **27**
- 3.1 Mixed Ligand 27
- 3.2 Conclusion 50
- References 51

4 Complexity Based on Ligand—Part 2 **57**
- 4.1 Polytopic Linkers 57
- 4.2 Multi-Heterotopic Ligands 63
- 4.3 Conclusion 66
- References 66

5 Complexity Based on Metal Node **71**
5.1 Mixed Metal 71
5.2 Multiple SBUs 81
5.3 Conclusion 92
References 92

6 Complexity Based on Chiral Framework—Part 1 **105**
6.1 Inherent Chirality 105
6.2 Direct Chirality 109
6.3 Conclusion 120
References 121

7 Complexity Based on Chiral Framework—Part 2 **127**
7.1 Chiral-Template Synthesis 127
7.2 Post-Synthesis 131
7.3 Conclusion 144
References 144

8 Complexity Based on Structural Defects **149**
8.1 Inherent Defect 149
8.2 Designed Defect 154
8.3 Conclusion 161
References 162

9 Complexity Based on Heterogeneous Pores **171**
9.1 Heterogeneous Pores 171
9.2 Conclusion 178
References 179

10 Complexity Based on Mixed MOFs **185**
10.1 Complex Mixed MOFs 185
10.2 Conclusion 192
References 193

Index **199**

List of Illustrations

Figure 1.1	Simple description of 3D MOF chemistry.	2
Figure 1.2	Some examples of metal nodes, organic linkers, and MOFs (definition of atom types: blue: metal; red: oxygen; purple: nitrogen; gray: carbon; and green: chlorine).	3
Figure 1.3	Some examples of MOFs synthesis methods.	7
Figure 2.1	Simple language to understand concept of complexity.	14
Figure 2.2	Overview of this book based on the effective factors in the construction MOFs with heterogeneous structures.	16
Figure 2.3	Classification of complexity key factors in MOFs (further details in text).	16
Figure 2.4	A cost-effective mixed-metal mixed-ligand MOF, which exhibits highly efficient photocatalytic H_2 generation.	19
Figure 2.5	Defective linker concept for defect-engineered MOFs.	20
Figure 3.1	A photocatalyst MOF with three different ditopic linkers.	33
Figure 3.2	Structural analyses of a MOF-based photocatalyst with three different ditopic linkers. (a) Percent of BPDC-$(HN_2)_2$ incorporation in ReMOF-NH_2 (X%). (b) PXRD patterns: Re-MOF, Re-MOF-NH_2 (33%), Re-MOF-NH_2 (80%) and Re-MOF simulated pattern. (c) SEM images: Re-MOF-NH_2 (33%) and Re-MOF-NH_2 (80%). (d) N_2 adsorption isotherms: Re-MOF, Re-MOF-NH_2 (33%) and Re-MOF-NH_2 (80%).	

(e) IR spectra: Re-MOF, Re-MOF-NH_2 (33%) and Re-MOF-NH_2 (80%). 34
Figure 3.3 (a) Copper-phosphonate CBU polyhedral. Octahedral Cu1 and square pyramidal Cu2: dark blue and light blue, respectively. PO_3C tetrahedral: green. (b) The original structure of the considered isoreticular MOF. 36
Figure 3.4 Schematic representations of the happened processes in MOF with increased porosity. (a and b) Organic linker installation and linker labilization, respectively. (c) The produced hierarchically porous MOF through linker labilization (top). UV-vis analysis of linker exchange process. (a) Concentration of AZDC in supernatant as a function of incubation time in CBAB solutions with different concentrations. (b) Relationship between CBAB exchanged/CBAB added, exchange ratio, and CBAB molarity. (c) PCN-160 crystals images with various exchange ratios. 38
Figure 3.5 Schematic representations of construction mechanism. (a) Formed micropores, (b and c) small and large mesopores, and (d) pores size distribution of the considered MOF (PCN-160-34%) in the presence of acid in different amounts. (Top) (a) Representation of one kind of MOF in the presence of enzyme. (b) relative activity of the considered MOFs in the oxidative reaction of ABTS (2,2′-azino-bis(3-ethylbenzothiazoline-6-sulphonic acid)) and o-PDA (o-phenylenediamine). 40
Figure 3.6 One kind of mixed-valence $Ru^{II/III}$ MOF with mixed-linker. 42
Figure 3.7 Olefin hydrogenation mechanism. 42
Figure 3.8 Zeolite-like MOFs based on mixed linkers. 44
Figure 3.9 A mixed-ligand MOF with two ligands BTC (benzene-1,3,5-tricarboxylate) and BTRE (1,2-bis(1,2,4-triazol-4-yl)-ethane). 44

Figure 3.10 Stepwise preparation of {[$Cu_4(CDC)_4$(4,4′-bipy)$(H_2O)_2]_3\}_n$·xS. Copper: aqua; oxygen: red; nitrogen: blue; carbon; black. Hydrogen atoms have not been shown (top). Synthesis of the isostructural porphyrinic MOFs and obtained crystals photographs (bottom). 46
Figure 3.11 ADES-1: (a) coordination environment around Zn(II); (b) and (c) $[Zn_2(COO)_2]$ SBU in 1D metal-carboxylate chain and double lined 2D network, respectively; (d) ABAB manner interlayer with lattice H_2O. 47
Figure 3.12 ADES-2: (a) coordination environment around Cd(II); (b) and (c) $[Cd_2(COO)_2]$ SBU in 1D metal-carboxylate chain and double lined 2D network, respectively; (d) Offset stacked 2D layers with the lattice H2O and π···π stacked L. 47
Figure 3.13 Direct synthesis mixed-ligands MOF film, RuB-RuTB-UiO-67/TiO_2/FTO through solvothermal method. 50
Figure 4.1 Some of the tritopic linkers. 58
Figure 4.2 The symmetry in organic linker. Tetratopic linkers with (a) symmetry $C2_h$ and (b) symmetry C_s. 59
Figure 4.3 Some of the tetratopic linkers. 60
Figure 4.4 A mixed-ligand MOF with tetratopic organic linkers. 61
Figure 4.5 Some of the multi-topic linkers. 62
Figure 4.6 A kind of MOF with multi-hetero topic organic linker. 64
Figure 4.7 (a) Cu(II) Coordination environment. Symmetric nodes: (i) y, 1 - x, 1 - z; (ii) y, x, 1 – z; (iii) 1 - y, x, 1 – z; (iv) 1 – x, y, z; (v) x, 1 – y, z; (vi) -x, 1 – y, z; (vii) -x, y, z; (viii) y, x, -z; (ix) -y, x, -z; (x) y, -x, -z. (b) Distorted truncated octahedron cage: (trinuclear + tetranuclear) SBUs. (c) Illustration of three dimensional porous framework in the considered MOF. 65
Figure 5.1 Two synthesis strategies to create MM′-MOFs. 72

Figure 5.2 Two synthetic methods to create 2D gird MM′-MOFs. 73
Figure 5.3 The incorporated BPDC, RuDCBPY, and PtDCBPY into Zr-MOF (UiO-67) through the solvothermal method. Red balls: SBUs. Gray rod: BPDC. Purple rod: RuDCBPY. Yellow rod: PtDCBPY. This figure is copied with the permission of the mentioned reference. 74
Figure 5.4 (a) Powder x-ray diffraction patterns, (b) IR spectra, and (c) TGA analysis of UIO-67_{black}, Pt@UIO-67_{red}, and Ru-Pt@UIO-67_{blue}. (d) Connolly-surface-filling model demonstrates mother MOFs of UIO-67 (pore size: 8–10 Å). 75
Figure 5.5 (a–c) SEM and (d–j) HAADF-STEM images of the considered samples (top). UV-vis of bpdc, RuDCBPY (aqueous solution), and PtDCBPY (MeOH). (a, b) Diffuse reflectance of the considered samples (middle). (c) Powdery samples photographs under natural light (bottom). 76
Figure 5.6 The formation of coordination polymers from the corresponding metals [4 to 8]/[Cu(2,2′-bpy)]$^{2+}$[9]/[10] and metalloligand [L^{Cu}]. 77
Figure 5.7 The epitaxial growth of the trimetallic and bimetallic hetero-(Ln)-MOFs from the crystal seed with one type of metal. 79
Figure 5.8 Transmetalation in porph@MOM-10 (a) and MSO-MOF (b). [Panel (b): metalation after demetalation of MOF]. 80
Figure 5.9 (a) 3D structure of $[Gd_3Cu_{12}I_{12}(IN)_9(DMF)_4]_n$.nDMF. (**structure I**) (b) $[Cu_{12}I_{12}]$ cluster. (c) Tri-nuclear $[Gd_3(IN)_9(DMF)_4]$ units (top). (a) 3D structure of $[Gd_4Cu_4I_3(CO_3)_2(IN)_9(HIN)_{0.5}(DMF)(H_2O)]_n$.nDMF.n$H_2O$. (**structure I**) (b) $[Cu_3I_2]$ cluster. (c) $Gd_4(CO_3)_2$ chains (bottom). 82
Figure 5.10 The solid-state luminescent properties of MOFs with diverse SBUs. 85
Figure 5.11 (a and b) MOF 1 structure showing Co_3 SBU supported by Co1 and Co2 viewing from two

directions. (c) Three-dimensional network viewing along the [010] direction. 86

Figure 6.1 (a) Two types of chiral tetrahedral SBUs. (b) A perspective view of microporous framework: zinc, azury; carbon, gray; oxygen, red. Water molecules of lattice are eliminate for clarity (left). Adsorption isotherm of methane at 298 K, and H_2 adsorption desorption isotherms (77 K, inset) activated at 100°C (blue) and 175°C (red) (right and top). Polarization (P) versus applied field (E) hysteresis loops from a single crystal sample (right and bottom). 108

Figure 6.2 (a) Schematic representation of a chiral MOF catalyst by PSM. (b) The chemical structures of the used chiral ligands based on tetracarboxylic acids. 110

Figure 6.3 Crystal structure of chiral MOF-1a. (a) Paddle-wheels clusters based on Cu and their connectivity with chiral L1a ligands. (b) L1a: blue distorted tetrahedro; Cu-paddle-wheel cluster: red square. (c) Connectivity of L1a organic ligands and Cu-paddle-wheel node. (d) Stick model. (e) Representation of a simplified connectivity (top). (a–h) Space-filling models of CMOF-1b to -4b with pores size (bottom). 112

Figure 6.4 (a and b) Views of connectivity around metallic centers in structure 2, highlighting D,L-enantiomeric Cd2+ ions. (c) The one-dimensional zigzag coordination chain structure in 2. Cyan: Cd(1); pink: Cd(2) atom; red: oxygen atom; blue: nitrogen atom; gray: carbon atom; yellow: hydrogen atom. α-methyl groups of D-HCam organic ligands: green-colored balls (similar to Zn-analogous 3). 117

Figure 6.5 Total synthesis of CMOF with chiral axial ligand. 118

Figure 6.6 Different cavity sizes in CMOFs. 118

Figure 6.7	Stick/polyhedra models of CMOF-1-5 with the presence of interpenetration along with connectivity of ligands and SBUs.	119
Figure 7.1	An example of the chiral templating effect origin.	128
Figure 7.2	Synthesis method of Ni-PYI1, guest exchange, and the considered asymmetric catalysis.	129
Figure 7.3	Examples of the post-chiral modification of achiral MOF-NH_2 as capillary columns for chiral GC.	132
Figure 7.4	(a–d) GC chromatograms based on chiral MOF-based columns to separate racemates under different conditions. (a) MIL-101-S-2-Ppa coated column A: 2-methyl-2,4-pentanediol (160°C, 2 ml min^{-1} N_2); 1,2-pentanediol (160°C, 2 ml min^{-1} N_2); citronellal (165°C, 2 ml min^{-1} N_2). (b) MIL-101-R-Epo coated column B: 2-butanol (150°C, 2 ml min^{-1} N_2); 1-heptyn-3-ol (200°C, 2 ml min^{-1} N_2); citronellal (180°C, 2 ml min^{-1} N_2). (c) MIL-101-(+)-Ac-L-Ta coated column C: 1-amino-2-propanol (210°C, 2 ml min^{-1} N_2); 2-amino-1-butanol (210°C, 2 ml min^{-1} N_2); 1,2-pentanediol (210°C, 3 ml min^{-1} N_2). (d) MIL-101-L-Pro coated column D: Mandelonitrile (200°C, 1.8 ml min^{-1} N_2); 1-phenylethylamine (180°C, 2 ml min^{-1} N_2); methyl-2-chloropropionate (180°C, 2 ml min^{-1} N_2).	133
Figure 7.5	The example of modified MOF with chiral species through post-chiral modification.	134
Figure 7.6	Asymmetry unit in Zn-DPYI crystal structure (left). Crystal structure of Zn-DPYI (right).	134
Figure 7.7	The preparation steps of one kind of chiral Zn-MOF through post-synthesis modification by using two chiral compounds with the opposite chirality.	138
Figure 8.1	A plot based on the number of published articles about the defect of MOFs.	152
Figure 8.2	MOF with interpenetration.	153

Figure 8.3 Different cases of MOFs with various mixed linkers. (a) The parent MOF, (b) the different mixed linker structurally, and (c and d) the large and truncated mixed organic linkers. 155
Figure 8.4 MIL to different forms: (a) without any defect, (b) dangling organic linker, and (c) linker vacancy. The orange polyhedral: cationic units. C atom: black. O atom: blue. 156
Figure 8.5 Defective UiO-66(Zr)-$(OH)_2$ framework for CO_2 capture. 161
Figure 9.1 (a) H_4PANAD structure. (b) HHU-1 network with varying size pores (two brown atoms illustrate one ellipsoid-like pore). For clarity, H atoms have not been shown. 172
Figure 9.2 Heterogeneous pores due to missing node or ligand. 174
Figure 9.3 An example of MOF-based porous carbon (NHOPC) with multiporosity synthesis as a supercapacitor. 176
Figure 9.4 Examples of Cr-MILs as mesoMOFs with two types of mesocages. 178
Figure 10.1 Schematic representation of bio-MOF composite. 187
Figure 10.2 Heteroepitaxially grown hybrid IRMOF-3@1 and structures of IRMOF-1 and IRMOF-3 (zinc: purple; nitrogen: blue; oxygen: red; carbon: gray. Hydrogen atoms have not been shown). 188
Figure 10.3 Optical micrographs: (a) hybrid IRMOF-3@1, (b) IRMOF-1@3. (IRMOF-1: transparent; IRMOF-3: brownish). 189
Figure 10.4 Isostructural Eu/Tb mixed MOFs with different triplet energy levels. 192

List of Tables

Table 3.1	Examples of mixed ligands metal-organic frameworks.	29
Table 3.2	Determination of catalytic conditions for Biginelli reaction.	49
Table 5.1	Optimization of the carboxylation of phenylacetylene.	83
Table 5.2	Carboxylation of terminal alkynes with CO_2.	84
Table 5.3	Crystallographic data and refinement parameters of the considered MOF.	87
Table 5.4	Various applications of mixed-metal MOFs.	88
Table 6.1	Diethylzinc additions to aromatic aldehydes.	113
Table 6.2	Alkynylzinc additions to aromatic aldehydes (similar conditions with Table 6.1).	114
Table 7.1	Asymmetric dihydroxylation of aryl olefins.	130
Table 7.2	Aldol reactions catalyzed by Zn-MOF1.	135
Table 7.3	Aldol reactions by using Zn-MOF1 and Zn-MOF2.	136
Table 7.4	Control reactions for asymmetric epoxidation of styrene.	140
Table 7.5	Asymmetric epoxidation of several olefins and methanolysis of styrene oxide.	142
Table 8.1	Some of the published papers about defects in metal-organic frameworks.	151
Table 8.2	Some of defective MOF-based materials.	157
Table 10.1	Carbon dioxide cycloaddition with various epoxides by using CZ-BDO.	190

List of Schemes

Scheme 2.1	The conversion of an achiral MOF to a chiral MOF by using an achiral organic linker.	19
Scheme 3.1	Structure of the used ligands.	35
Scheme 3.2	Structures of the used organic dyes.	48
Scheme 3.3	Biginelli reaction by using ADES-1.	48
Scheme 6.1	Synthesis conditions to produce chiral metal-camphorate coordination polymers 1–5.	115
Scheme 6.2	Coordination modes of D-Cam and D-HCam in polymeric structures 1–5.	116
Scheme 7.1	Synthesis method of chiral [MIL-101(Cr)-tart].	137
Scheme 7.2	Interactions in the preparation of [MIL-101(Cr)-tart].	139
Scheme 7.3	The preparation steps of post-chiral modification NU-1000 by SALI method and then Mo-catalyst immobilization on chiral species.	143
Scheme 10.1	Suggested mechanism of carbon dioxide cycloaddition with epoxide by bimetal mixed MOFs.	191

Preface

The study of metal-organic frameworks (MOFs) as porous crystalline materials has risen rapidly over recent years due to their unique properties and various potential applications. Until now, thousands of MOFs have been prepared with wonderful new designs and further precision in the synthesis methods.

As known, MOFs chemistry is started with the main two components: inorganic and organic units. But, it is of interest to know that most of them have homogeneous structure with having one of each kind of component. Recently, MOFs with heterogeneous structures have been attracting strong attention due to their interesting structure and features. These kinds of materials in their body have more kinds of components that lead to structural complexities. The understanding of effective agents in the production of MOFs with heterogeneous structures is another key and important concept that is not easy. Given that how these kinds of heterogeneous porous MOFs are constructed is attractive, so, in this book, we focus on advances in research on MOFs with heterogeneous structures covering mixed components (metals/ligands), multi-heterotopic linkers, defect, heterogeneous pores, mixed MOFs, and, most importantly, chiral MOFs along with some examples. These different cases lead to diversity of parent architectures, and finally, the complexity and heterogeneity appear in them. Many reports have been presented about different kinds of MOFs but some of them are very complex therefore the attempt to their understanding should be increased. This book can help further understanding of these kinds of MOFs.

Ali Morsali and Kayhaneh Berijani
Department of Chemistry, Tarbiat Modares University
Tehran, Islamic Republic of Iran
July 2021

Abbreviations

1,3-BDC	(1,3)-Benzene dicarboxylate
MUF	Massey University Framework
BTB	Benzene-1,3,5-tribenzoate
BTC	Benzene-1,3,5-tricarboxylate
1,4-BDC	1,4-benzenedicarboxylate
BPDC	4,4′-biphenyldicarboxylate
Bz-IM	Benzimidazole
OHC-IM	Carboxaldehyde-2-imidazole
D-Cam	D-camphanic acid
bpdc	4,4′-Diphenyl dicarboxylate
D-POST-1	$[Zn_3(\mu_3\text{-}O)(1\text{-}H)_6].2H_3O.12H_2O$
MM′-MOF	Mixed-metals MOF
MBBs	Molecular building blocks
MTV-MOF	Multivariate metal–organic framework
2-MeIM	2-Methylimidazole
NLO	Non-linear optics
PL-MOFs	Pillared layer metal-organic frameworks
PSM	Post-synthetic modification
SBUs	Secondary building units
SBBs	Supramolecular building blocks
SBLs	Supramolecular building layers
TBUs	Tertiary building units
F_5CPP	5,10,15-Tris(p-carboxyphenyl)-20 (pentafluorophenyl) porphyrin
ZABUs	Zinc-adeninate octahedral building units
ZJU	Zheijang University
H4TBAPy	1,3,6,8-Tetrakis(p-benzoic acid)pyrene
Zn-TCPP	[5,10,15,20-Tetrakis(4-carboxy-phenyl) porphyrinato]-Zn(II)

ICP-OES	Inductively coupled plasma-optical emission spectroscopy
Zn-MI	$[Zn_4O(3\text{-Methyl-5-isopropyl-4-carboxypyrazolate})_3]_n$
Zn-DM	$[Zn_4O(3,5\text{-Dimethyl-4-carboxypyrazolate})_3]_n$
SAM	Self-assembled monolayer
Dcam	(1R,3S)-(+)-Camphoric acid
dabco	Diazabicyclo(2.2.2)octane)
ndc	Naphthalene dicarboxylate
dabco	1,4-Diazabicyclo[2.2.2]octane)
bipy	Bipyridine
bpe	Bis(4-pyridyl)-ethylene
dpndi	N,N'-Di(4-pyridyl)-1,4,5,8-naphthalene tetracarboxydiimide
adc	Anthracene dicarboxylate
NDC	2,6-Naphthalenedicarboxylate
RT	Room temperature
DHPMs	3,4-dihydropyrimidin-2(1H)-ones
FTO	Fluorine-doped tin oxide

1

Introduction: A Brief Introduction About Metal-Organic Frameworks

Abstract
Metal-organic frameworks (MOFs) or porous coordination polymers (PCPs) are a subset of hybrid porous materials that have recently attracted considerable attention. Their structures are constructed from inorganic (such as metal clusters or atoms, rod-shaped clusters) and organic (such as carboxylates, azoles, nitriles) parts with different chemical nature. The features of both parts (node and linker) can determine MOF properties like network structure (net) topology, physical, mechanical and morphological features. MOFs with tunable pore size, different functionalization, high surface area and stability can be synthesized through the various synthesis methods. Depending on application of MOFs, design and their synthetic methods can be changed that are very important. Indeed, the rise of MOFs chemistry can be primarily related to their properties and performances. In this chapter, we present a brief introduction about MOFs such as types of metal nodes, organic linkers and synthesis strategies. So, based on the used components in MOFs, diverse frameworks can be designed and generated. As a result, tunability in their structure makes them attractive and important candidates for diverse purposes.

Keywords: Metal-organic framework, inorganic building units, organic linkers, structural tunability, porous materials

1.1 Metal-Organic Frameworks

Over almost 20 years, many investigations have been performed about the porous materials field especially metal-organic frameworks (MOFs) as a subset of porous coordination polymers (PCPs). For the

Ali Morsali and Kayhaneh Berijani. Metal-Organic Frameworks with Heterogeneous Structures, (1–12) © 2021 Scrivener Publishing LLC

first time, these compounds were defined by Yaghi and co-workers in 1995. The considerable attention to MOFs is due to their chemical and physical properties such as variety in building units and design, pore size, and cavity shape that lead to different frameworks with diverse applications. In the growth of these compounds and their introduction, several scientific groups are trying. The most important and famous MOFs are MOF-5 (zinc salt and H_2BDC), HKUST-1 (copper salt and H_3BTC), MIL-101 (chromium/iron salt and H_2BDC), and MOF-74 (zinc salt and H_2-DHBDC). As known, the reticular structures of MOFs can be made from two sections: inorganic nodes and organic linkers (sometimes with functional group). Simple description of 3D MOF chemistry has been shown in Figure 1.1.

Diverse building blocks can be used in the synthesis of any MOF, for example, metal units as node including lanthanide cations, transition metals, heavy transition metals and main metal ions, organic linkers like carboxylates linkers, and N-donor ligands that can be ditopic or polytopic. So, unlimited number of frameworks can be created with different features, geometries, and structures. Whereas, these components as building blocks can affect MOFs properties, so they should be carefully selected. In the following, we show some examples of SBUs (secondary building units), linkers, and MOFs although there are the various building units (Figure 1.2) [1, 2].

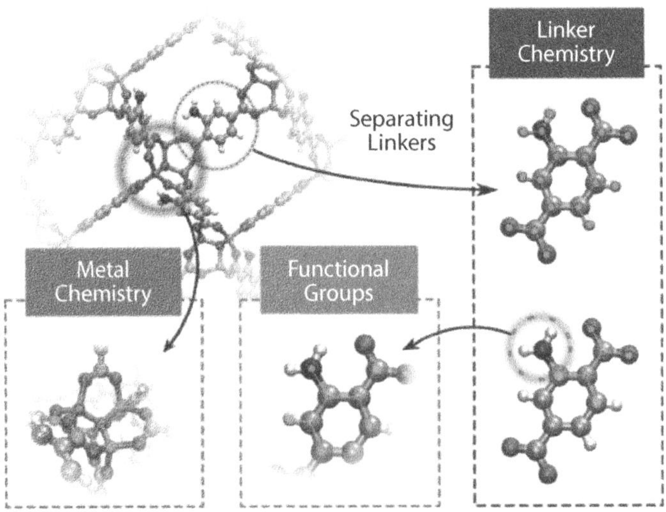

Figure 1.1 Simple description of 3D MOF chemistry [1].

It is worthy to note that stability and instability of MOFs are closely related to the used components and their coordination bonding (between metal moiety and organic linker). In the construction of these PCPs with crystalline networks, the coordination bonds are formed between building units. They have desirable features such as large surface area, adjustable pores, appropriate porosity, and crystalline structure with different topologies [3–5]. In addition, the body of MOFs, from metal nodes to linkers, can be functionalized with various functional groups through different strategies such as incorporation of the intended groups by post-synthesis modification (PSM) as straightforward approach. The ideal tunability in MOFs constituents leads to various platforms with different applications

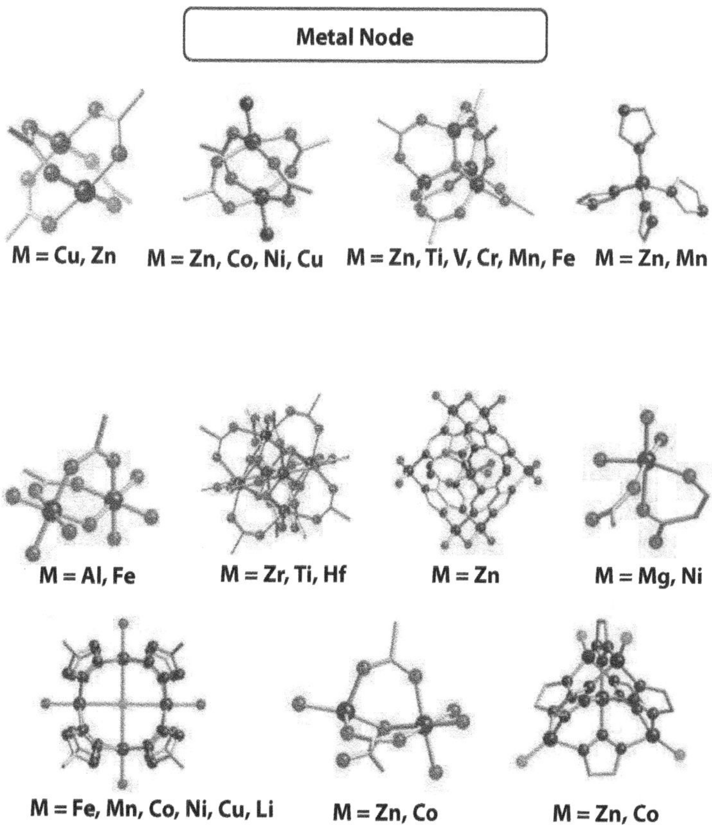

Figure 1.2 Some examples of metal nodes, organic linkers, and MOFs (definition of atom types: blue: metal; red: oxygen; purple: nitrogen; gray: carbon; and green: chlorine). *(Continued)*

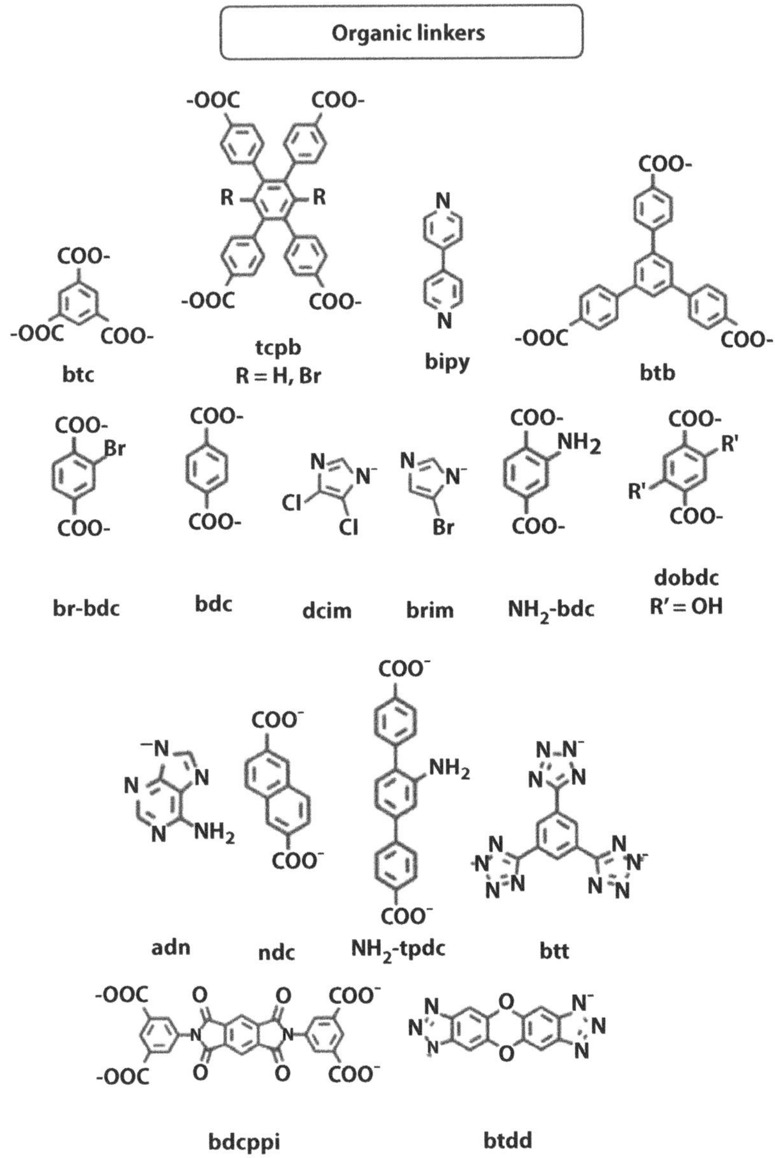

Figure 1.2 (Continued) Some examples of metal nodes, organic linkers, and MOFs (definition of atom types: blue: metal; red: oxygen; purple: nitrogen; gray: carbon; and green: chlorine). *(Continued)*

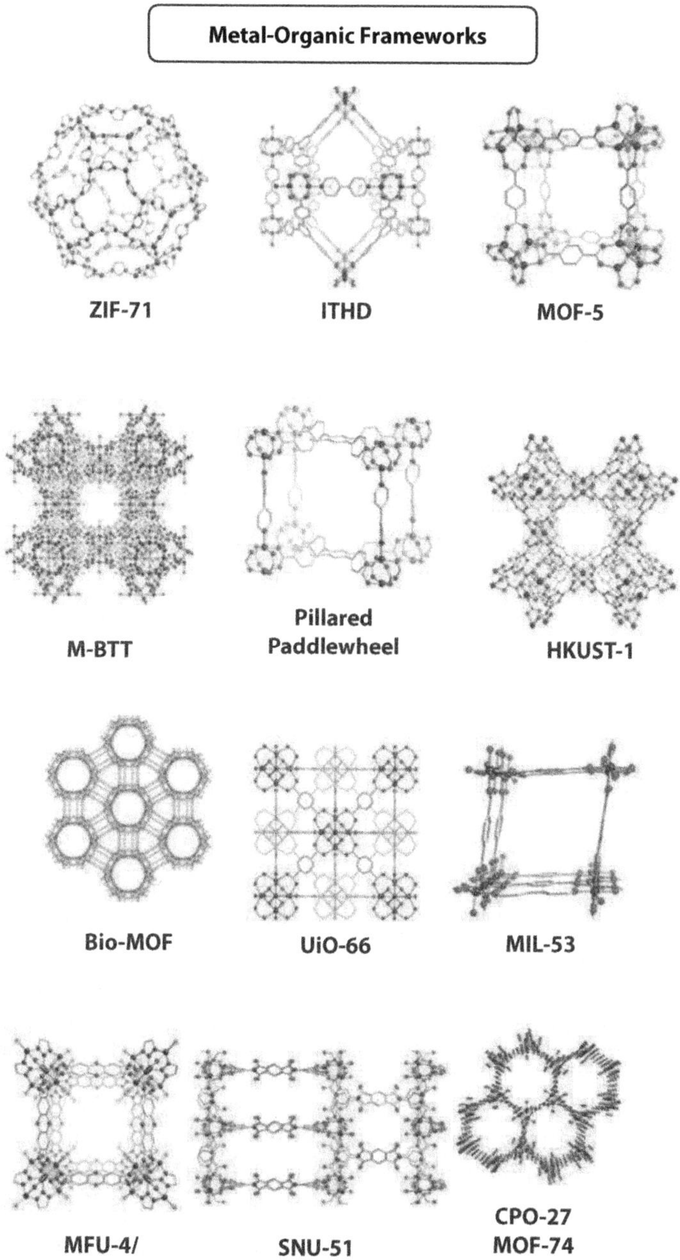

Figure 1.2 (Continued) Some examples of metal nodes, organic linkers, and MOFs (definition of atom types: blue: metal; red: oxygen; purple: nitrogen; gray: carbon; and green: chlorine).

even in the industrial scale as catalysis, gas storage, supercapacitors, medical uses, separation, sensing, and so on. In fact, owing to the facile ability to functionalize such materials, MOFs have found suitability for a range of applications. Certainly, there are some difficulties and limitations to prepare the plausible nets of MOFs such as temperature, solvent, stability, geometry building blocks, kind of the used metal/ligand, and their solubility. Nowadays, new ideas and valuable efforts have been aimed to the successful synthesis of new MOFs with remarkable designs and useful properties, because MOFs have the desirable potential due to the flexibility in choosing the structural components. Until now, different approaches to synthesize porous MOFs have been generally reported such as solvothermal or sonochemical syntheses. But, day to day, the development in new or optimized synthetic methodologies accelerates the creation of new structures with different features. As an example, Hwang *et al.* reported the different synthetic methods in Fe-BTC hydrothermal synthesis based on change in time or temperature reaction, or Yang *et al.* presented conditions of hydrothermal synthesis in the generation of HKUST-1 based on change in time and solvent. Some synthesis methods have been introduced in Figure 1.3. In all of them, the optimal conditions of synthesis methods have been investigated with their advantages and disadvantages. However, MOFs can be easily manipulated and modified by the adjustment of the components and synthesis conditions [6–15]. Indeed, their consistuents enable the preparation of different frameworks with various sizes through a broad spectrum of synthetic methods like solvothermal (such as hydrothermal synthesis: it refers to synthesis through chemical reactions in an aqueous solution above the boiling point of water), electrochemical method (direct or indirect synthesis; Unlike indirect synthesis, in which electrochemistry is only a step in an overall synthesis method, direct route allows synthesis rate control by electrochemical means) [16–19], sonochemical synthesis (using powerful ultrasound radiation (20 kHz–10 MHz)), [20, 21] mechanochemical synthesis (a processing method of solids in which mechanical and chemical phenomena are coupled on a molecular scale) and so on [22–24]. As mentioned, depending on the synthesis conditions and compositions, properties of MOF can be changed toward new and modified features. So, structural flexibility and diversity of MOFs that

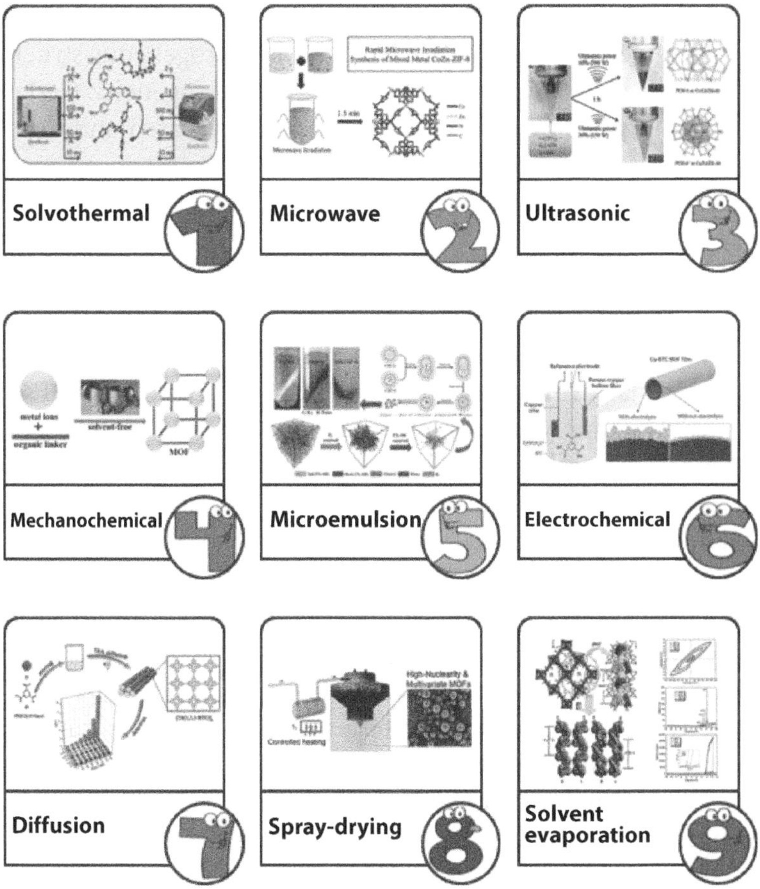

Figure 1.3 Some examples of MOFs synthesis methods.

can be originated from synthetic methods can promote and improve frameworks and their applications. For example, the combination of 1,4-benzenedicarboxylate and Zn leads to 1D linear structure. It is interesting that 2D layered structure and 3D network can be also produced by using these starting materials through interactions such as hydrogen bond and π-π stacking [25, 26]. So, due to a great number of metal nodes and organic linkers, designed new well-defined MOFs through different syntheses can be created. Most porous MOFs have homogeneous structures without any heterogeneity or complexity in their structure. The term of "complexity" in materials has received a great deal of attention in scientific fields. However, complexity in systems has a long history in science. In this book, we introduce some effective factors in the production of heterogeneous structures of MOFs by their structural complexity and heterogeneity. Certainly, book chapters can provide creativity for new research and describe a number of significant and different up-to-date strategies about the generation of complex MOFs.

1.2 Conclusion

Until now, many investigations have been performed in MOFs field from synthesis approaches to analytical characterizations along with theoretical calculations. According to the capabilities of MOFs and their derived materials, investigation and use of them have been increased in the different fields such as optical and electrochemical sensors, catalysis, gas storage/separation, drug delivery, and ion-conduction. Indeed, the rise of chemistry about MOFs can be ascribed to the relationship between physical and chemical features of MOFs with their structure/composition like pore size, surface area, active sites, stability, and interaction degree. Most frameworks based on MOF are in homogeneous form, but recently, considerable efforts have been devoted to the creation of robust, stable MOFs with heterogeneous structures. So, establishment of several rational strategies for generating complex MOFs with developing new process conditions has attracted much more attention. In relation to MOFs with heterogeneous structures, the next chapters discuss the introduction of the kinds of complexities and their progresses in MOFs.

References

1. Moosavi, S.M., Nandy, A., Jablonka, K.M., Ongari, D., Janet, J.P., Boyd, P.G., Lee, Y., Smit, B., Kulik, H., Understanding the diversity of the metal-organic framework ecosystem. *Nat. Commun.*, *11*, 4068, 2020.
2. Furukawa, H., Cordova, K.E., O'Keeffe, M., Yaghi, O.M., The chemistry and applications of metal-organic frameworks. *Science*, *341*, 6149, 1230444, 2013.

3. Furukawa, H., Ko, N., Go, Y.B., Aratani, N., Choi, S.B., Choi, E., Yazaydin, A.O., Snurr, R.Q., O'Keeffe, M., Kim, J., Yaghi, O.M., Ultrahigh porosity in metal-organic frameworks. *Science, 329*, 424–428, 2010.
4. Chen, S.S., Hu, C., Liu, C.H., Chen, Y.H., Ahamad, T., Alshehri, S.M., Huang, P.H., Wu, K.C.W., De Novo synthesis of platinum-nanoparticle-encapsulated UiO-66-NH_2 for photocatalytic thin film fabrication with enhanced performance of phenol degradation. *J. Hazard. Mater., 397*, 122431, 2020.
5. Ghasempour, H. and Morsali, A., Function-Topology Relationship in Catalytic Hydrolysis of Chemical Warfare Simulant inTwo Zr-MOFs. *Chem.-Eur. J*, 2020, *26*, 17437–17444, 2020.
6. Berijani, K., Morsali, A., Hupp, J.T., An Effective Strategy for Creating Asymmetric MOFs for Chirality Induction: A Chiral Zr-based MOF for Enantioselective Epoxidation. *Catal. Sci. Technol., 9*, 3388–3397, 2019.
7. Ghasempour, H., Wang, K.Y., Powell, J.A., ZareKarizi, F., Lv, X.L., Morsali, A., Zhou, H.C., Metal–organic frameworks based on multi-carboxylate linkers. *Coord. Chem. Rev., 426*, 213542, 2021.
8. Macreadie, L.K., Babarao, R., Setter, C.J., Lee, S.J., Qazvini, O.T., Seeber, A.J., Tsanaktsidis, J., Telfer, S.G., Batten, S.R., Hill, M.R., Enhancing Multicomponent Metal–Organic Frameworks for Low Pressure Liquid Organic Hydrogen Carrier Separations. *Angew. Chem. Int. Ed., 59*, 15, 6090–6098, 2020.
9. Razavi, S.A.A., Berijani, K., Morsali, A., Hybrid nanomaterials for asymmetric purposes: green enantioselective C–C bond formation by chiralization and multi-functionalization approaches. *Catal. Sci. Technol.*, 2020, *10*, 8240–8253, 2020.
10. Li, Z., Liu, P., Ou, C., Dong, X., Porous Metal–Organic Frameworks for Carbon Dioxide Adsorption and Separation at Low Pressure. *ACS Sustain. Chem. Eng., 8*, 41, 15378–15404, 2020.
11. Valekar, A.H., Lee, M., Yoon, J.W., Kwak, J., Hong, D.Y., Oh, K.R., Cha, G.Y., Kwon, Y.U., Jung, J., Chang, J.S., Hwang, Y.K., Catalytic transfer hydrogenation of furfural to furfuryl alcohol under mild conditions over Zr-MOFs: Exploring the role of metal node coordination and modification. *ACS Catal., 10*, 6, 3720–3732, 2020.
12. Zhang, M., Ding, X., Zhan, Y., Wang, Y., Wang, X., Improving the flame retardancy of poly (lactic acid) using an efficient ternary hybrid flame retardant by dual modification of graphene oxide with phenylphosphinic acid and nano MOFs. *J. Hazard. Mater., 384*, 121260, 2020.

13. Kalaj, M. and Cohen, S.M., Postsynthetic Modification: An Enabling Technology for the Advancement of Metal–Organic Frameworks. *ACS Cent. Sci.*, *6*, 7, 1046–1057, 2020.
14. Xu, C., Ai, Y., Zheng, T., Wang, C., Acoustic manipulation of breathing MOFs particles for self-folding composite films preparation. *Sens. Actuators A: Phys.*, *315*, 112288, 2020.
15. Peng, J., Li, Y., Sun, X., Huang, C., Jin, J., Wang, J., Chen, J., Controlled Manipulation of MOFs Layers to Nanometer Precision inside large Meso-channels of Ordered Mesoporous Silica for Enhanced Removal of Bisphenol A from Water. *ACS Appl. Mater. Interfaces*, *11*, 4, 4328–4337, 2019.
16. Al-Kutubi, H., Gascon, J., Sudhölter, E.J., Rassaei, L., Electrosynthesis of metal–organic frameworks: challenges and opportunities. *ChemElectroChem*, *2*, 4, 462–474, 2015.
17. Lestari, W.W., Winarni, I.D., Rahmawati, F., Electrosynthesis of Metal-Organic Frameworks (MOFs) Based on Nickel (II) and Benzene 1, 3, 5-Tri Carboxylic Acid (H_3BTC): An Optimization Reaction Condition. *In IOP Conf. Ser.: Mater. Sci. Eng.*, *172*, 1, 012064, 2017.
18. Ji, L., Hao, J., Wu, K., Yang, N., Potential-Tunable Metal–Organic Frameworks: Electrosynthesis, Properties, and Applications for Sensing of Organic Molecules. *J. Phys. Chem. C*, *123*, 4, 2248–2255, 2019.
19. Kumar, R.S., Kumar, S.S., Kulandainathan, M.A., Efficient electrosynthesis of highly active $Cu_3(BTC)_2$-MOF and its catalytic application to chemical reduction. *Micropor. Mesopor. Mater.*, *168*, 57–64, 2013.
20. Yu, K., Lee, Y.R., Seo, J.Y., Baek, K.Y., Chung, Y.M., Ahn, W.S., Sonochemical synthesis of Zr-based porphyrinic MOF-525 and MOF-545: Enhancement in catalytic and adsorption properties. *Micropor. Mesopor. Mater.*, *316*, 110985, 2021.
21. Wiwasuku, T., Othong, J., Boonmak, J., Ervithayasuporn, V., Youngme, S., Sonochemical synthesis of microscale Zn(ii)-MOF with dual Lewis basic sites for fluorescent turn-on detection of Al^{3+} and methanol with low detection limits. *Dalton Trans.*, *49*, 29, 10240–10249, 2020.
22. Chen, D., Zhao, J., Zhang, P., Dai, S., Mechanochemical synthesis of metal–organic frameworks. *Polyhedron*, *162*, 59–64, 2019.
23. Klimakow, M., Klobes, P., Thünemann, A.F., Rademann, K., Emmerling, F., Mechanochemical synthesis of metal–organic frameworks: a fast and facile approach toward quantitative yields and high specific surface areas. *Chem. Mater.*, *22*, 18, 5216–5221, 2010.

24. Singh, N.K., Hardi, M., Balema, V.P., Mechanochemical synthesis of an yttrium based metal-organic framework. *Chem. Commun.*, *49*, 10, 972–974, 2013.
25. Rodrigues, M.O., de Paula, M.V., Wanderley, K.A., Vasconcelos, I.B., Alves Jr, S., Soares, T.A., Metal organic frameworks for drug delivery and environmental remediation: A molecular docking approach. *Int. J. Quantum Chem.*, *112*, 20, 3346–3355, 2012.
26. Suh, M.P., Park, H.J., Prasad, T.K., Lim, D.-W., Hydrogen storage in metal-organic frameworks. *Chem. Rev.*, *112*, 782–835, 2012.

2
Metal-Organic Frameworks Complexity

Abstract

Until now, many materials have been synthesized, porous or non-porous that are simple or complex. Recently, the term of "complexity" as an important concept in materials has been introduced that focuses on the effective factors. Complex systems are the start of different perspectives that absolutely are related to design materials. About MOFs, we know that most of them are synthesized by using one kind of node and organic linker with homogeneous structure. But surprisingly, MOFs are created by multiple different kinds of building units with heterogeneous structure. In this chapter, we outline different strategies for production of complexity in MOFs based on the classification of complexity key factors. Depending on the kind of complexity grown, different types of MOFs are introduced.

Keywords: Complexity, multiple building units, heterogeneous structure, complex MOFs, heterogeneity

2.1 Perspectives on Complexity in MOFs

Microporous materials are significant in a wide range of applications, and they can exhibit particular properties such as high physical surface areas and the pores with different dimensions that can compete against others. There are several well-known microporous materials such as zeolites, silica, activated carbons, and metal-organic frameworks (MOFs). Until now, many MOFs as porous materials have been synthesized by using the various inorganic and organic species to form various frameworks. Diversity of metal sources and organic linkers lead to MOFs with specific crystallographic building, texture, and chemistry. Most of the reported structures are simple

Ali Morsali and Kayhaneh Berijani. Metal-Organic Frameworks with Heterogeneous Structures, (13–26) © 2021 Scrivener Publishing LLC

homogeneous forms with one kind of building units (linker and node), but recently, the synthetic MOFs have been made with multiple different components. Using the various building units in the preparation of MOFs can generate heterogeneous structures along with complexity. It is notable that heterogeneity in MOFs can cause structural complexity with the generation of different properties and applications. Here, we introduce and expand the variety of the heterogeneities which produce the complexity in MOFs such as complexity based on building blocks (mixed/polytopic/multiheterotopic ligands, mixed metal nodes, and multiple SBUs), frameworks (chiral frameworks, structural defects, and heterogeneous pores), and mixed MOFs. Indeed, heterogeneity in the materials is an important factor in the basic characterization and it can be generated during synthesis or induced into the structure by incorporation of the special components. For easy understanding of complexity term, the simple categories have been shown in Figure 2.1.

There are many books about MOFs, but still there is no book that covers whole of the aspects in the production of MOFs with heterogeneous structures. So, we discuss about the concept of it in MOFs

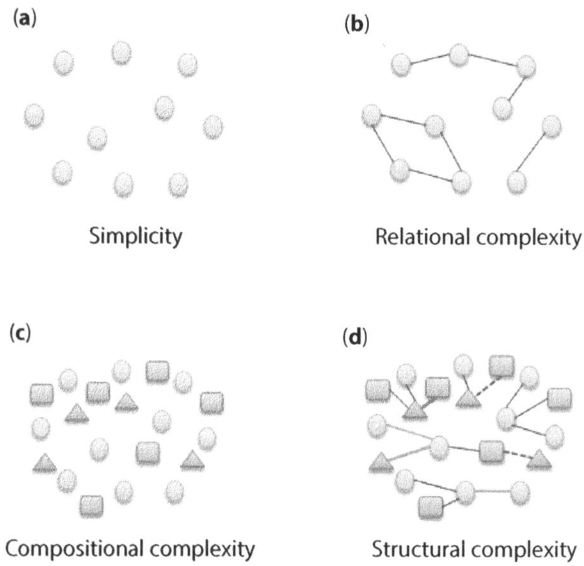

Figure 2.1 Simple language to understand concept of complexity [45].

with details and we hope that this book plays a special role for further understanding of complex MOFs.

Apart from homogeneous MOFs, recently, the construction of MOFs with heterogeneous structures along with complexity has been attracted. But, why, when, and how can complexity happen? These are questions for researchers, especially those who work in the MOFs field. To prepare this kind of MOFs, using the mixed components is expanded since they can induce heterogeneity to MOFs structure. Generally, all the above-mentioned factors can convert the simple MOFs with homogeneous state into heterogeneous structures. Given that no book has been made available about MOFs heterogeneous structures so, in this book, after some of the fundamental explanations, we focused on introducing the number of effective factors in MOFs complexity, design, synthesis procedures, function of these complex frameworks, and the comparison of their structural features to their homogeneous counterparts that surely they will be useful for scientists. So, please stay with us in the following book.

In 2015, Yaghi *et al.* reported a short review about varying building units within MOFs that led to frameworks with complexity[1]. Given that, these agents lead to MOFs with heterogeneous structures that have various applications, so we decided to further investigate in this field, namely MOFs with heterogeneous structures, in details. In Figure 2.2, overview of this book has been shown that it is based on the effective factors in the construction of MOFs with heterogeneous structures. Also, Figure 2.3 shows classification of key factors in the creation of these kinds of MOFs, in details. This classification is based on building blocks, framework and mixed-MOFs that can produce MOF-based heterogeneous structures. For enough understanding, the further explanations along with the related examples have been reported, as follows.

Many simple MOFs with different structures or three-dimensional coordination polymers (PCPs) were synthesized with well-known components through usual techniques as solvothermal, electrochemical, microwave, and ultrasonic-assisted synthesis [1, 2]. A wide range of these simple MOFs were prepared with single metal (metal ion or cluster) as node and organic linker as strut. The synthesis of

[1]*Angewandte Chemie International Edition, 54*(11), pp. 3417–3430.

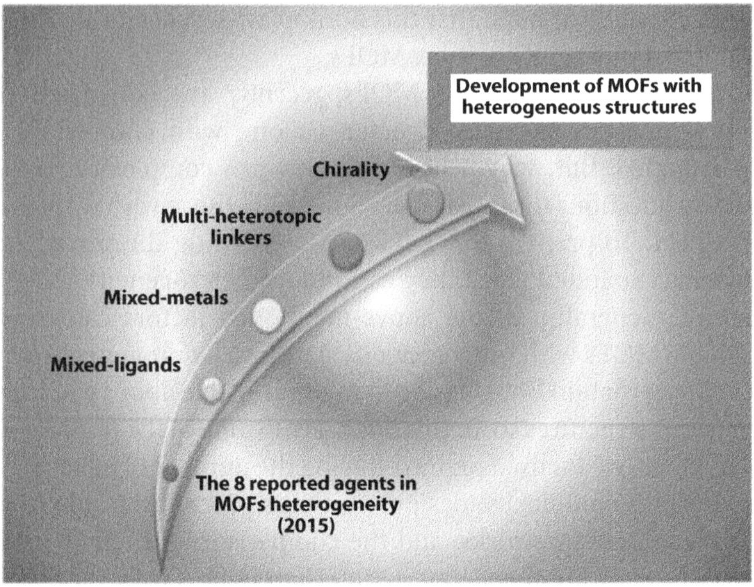

Figure 2.2 Overview of this book based on the effective factors in the construction MOFs with heterogeneous structures.

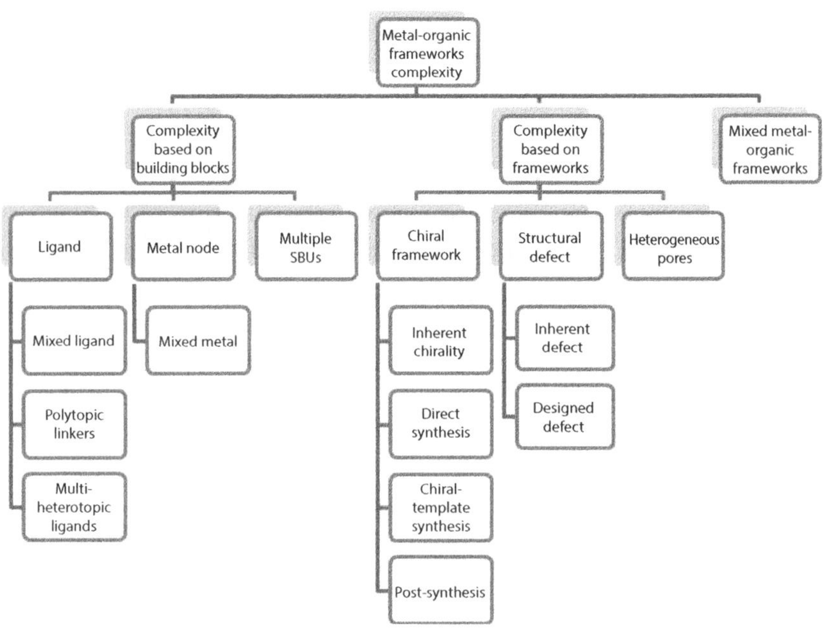

Figure 2.3 Classification of complexity key factors in MOFs (further details in text).

MOFs promoted with the diverse functional groups like NH_2, NO_2, Cl, F, R, and OR (R: organic chain). Day-to-day efforts to the designing of new MOFs increase with the distinctive features. According to the above discussion, MOFs due to their versatile structures can be introduced as flexible materials. Today, MOFs are constructed by using more than one kind of building blocks, linkers (mixed-linkers, polytopic linkers, and multi-heterotopic linkers), metal nodes (mixed-metals), and SBUs (secondary building units) that can produce heterogeneity which causes complexity in MOFs.

The generated complexity in MOF structure can also impress on MOFs properties and applications. Complexity concept means that a series of changes cause the homogeneous structure of MOFs to be converted to their heterogeneous counterparts. The existence of multiple pores in MOFs, chirality, defects, and disorders are the other effective factors in the creation of complexity in MOFs. These chemical factors cause that MOF crystalline structure to change rather than its perfect analogue. It is found that complexity in MOFs structure can be produced by the changing in the components and/or the addition of the various chemical compounds that have basic role in the design of final framework. High tunability in the chemistry of MOFs is one of the advantages that can provide the conditions for appearing the complexity. For example, the interweaving heterochiral helices in chiral MOFs, the heterogeneity in MOFs topology that can lead to the symmetry reduction of framework so complexity can increase. In some of the generated reticules from more than one kind of metal ions or linkers and or both of them, complexity can be observed due to different types of SBUs. Complexity in MOFs does not mean that the crystal structures of MOFs are damaged or their features lost but they can be desirable in the different forms with the various capabilities such as the surface defects in MOFs that can be employed as catalytic active centers in the catalytic processes so they are not always detrimental. In fact, the variation in MOFs components generates the new opportunities with different abilities. The clear classification of MOFs complexity has been done in the following and further details have been explained in the next chapters.

The variety of used components in the construction of MOFs is a very important point but there are some complexities. As mentioned, most of MOFs were prepared by linking of one kind of metal salts and linkers by strong bonds that are fundamental principles in MOFs chemistry. Surely, these properties affect the nature of material such as porosity, pore size, pore surface, and other physical properties [3].

At present, we confront the new MOFs with exciting crystal structures. MOFs can be synthesized by two or more different metal ions (mixed metals) and linkers (mixed linkers) that lead to the construction of the different SBUs. Naturally, SBUs as an essential concept in the production of MOFs are important organizing sections that their changing or variety can generate complexity in MOFs framework. The exchange of the parent components as linker can also create complexity in MOFs. This process in MOF framework can be assigned to BBR (building block replacement) that shows more interesting degree of structural complexity [4–7]. In this case, the decreasing symmetry in the used components leads to the coordination polymers with the increased complexity in their building blocks [8–10]. It can be said that the ligand exchange is a powerful post-synthetic approach that permits the diversity and complexity to increase in the MOF structures [11–16].

One kind of MOFs advantages is the presence of mixed metals in their creation. Almost, in our experiences, metal nodes of MOFs are single type, but recently, the several examples have been reported that metal section is mixed metals. Even, MOFs can be synthesized with mixed metals and mixed ligands, simultaneously. For example, a microporous MOF was prepared with mixed metals Na/Cu and flexible/rigid mixed ligands (4,4'-(4-(4H-1,2,4-triazol-4-yl)phenylazanediyl) dibenzoic acid/citric acid) with two kinds of channels quadrilateral and triangular [17]. Its ability was compared to parent MOFs that have single metal ion or single organic linker. The investigations showed the effective activity of mixed metal mixed ligand MOF in the photocatalytic system of H_2 generation is higher than homogeneous counterparts (Figure 2.4). Further studies on the framework of MOFs demonstrate that the chirality as an attractive property can also produce the complexity in MOFs. The interesting observations showed that both conditions (i) the existence of chiral agent and (ii) the components asymmetric arrangement in space can affect in the creation of complexity in MOFs [18, 19].

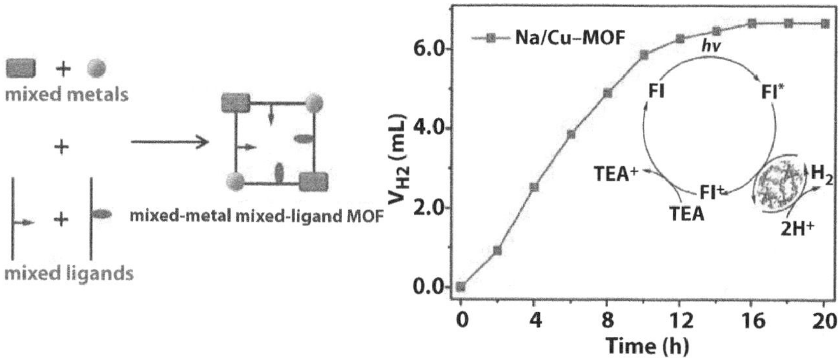

Figure 2.4 A cost-effective mixed-metal mixed-ligand MOF, which exhibits highly efficient photocatalytic H_2 generation [17].

For instance, an achiral MOF with adding an achiral precursor to structure can be converted to a chiral MOF due to the produced chiral axis (Scheme 2.1). Certainly, this phenomenon can generate the heterogeneity and cause MOF to exit from the homogeneous structure. Further explanations have been given in the related section. The heterogeneity in MOFs can be generated by many different changes in the frameworks such as the presence of different kinds of pores, vacancies, and defects. One of the examples that is said about the structural complexity in MOFs is the adding defective linkers to MOF framework during the preparing that can convert the defect free MOF to MOF with the considerable defect degree. Also, the missing organic ligands can cause the defects in framework of MOF [20–23].

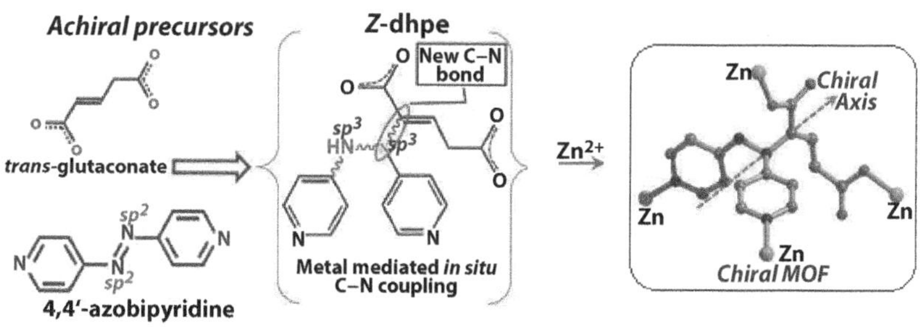

Scheme 2.1 The conversion of an achiral MOF to a chiral MOF by using an achiral organic linker [19].

In this line, it should be noted that along with the ligand doping, the related defects to mixed metals can be also created (mixed valence metals: Cu^{1+}/Cu^{2+}) [24]. In 2015, Yaghi *et al.* described heterogeneity within MOF order, although to us, the present review is nearly complete and ideal about the complexity and it differs from theirs [11].

Furthermore, the designing ligands, the length, and kind of their functional groups can affect the homogeneity of MOFs lattice (Figure 2.5). This figure is related to the defective linker doping into framework. In the parent framework, two Cu^{2+} sites at the axial positions of nodes have been identified. The used ligands are based on benzene-1,3-dicarboxylates such as L1: 5-nitroisophthalate, L2: 5-cyanoisophthalate, L3: 5-hydroxyisophthalate, L4: pyridine-3,5-dicarboxylate that their 5-position has different functional groups. The increasing defect degree is related to Lx^{2-} incorporation, for example btc^{3-}/Lx^{2-} exchange can happen that x in Lx can be 1 to 4. This process can produce opportunity for the creation of engineered defects on CUS and porosity that can be effective in the improvement of MOFs structure. Or a series of isoreticular bio-MOFs were synthesized by the mono-, di-, and tri-functionalization and the different functional groups as formyl, azido, and amino that can affect homogeneity of MOF structure [25]. Interestingly, increasing the size and complexity in the used linkers create better conditions to produce meso-MOFs (MOFs with increased porosity by using long and branched organic linkers). These kinds of linkers can generate problems in the characterization and determination of the framework structure of new MOFs [26, 27]. Omar M. Yaghi *et al.* described the

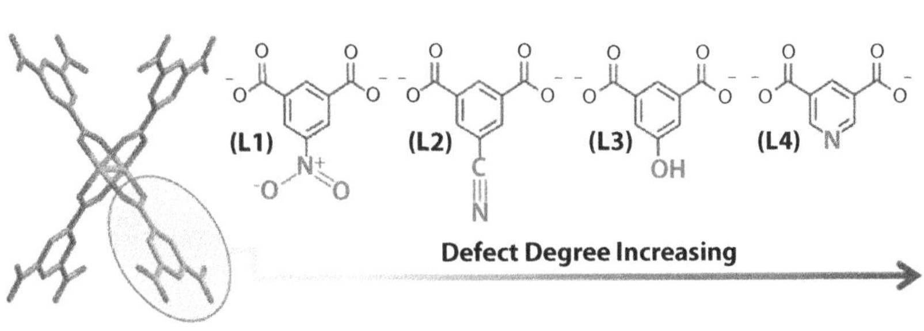

Figure 2.5 Defective linker concept for defect-engineered MOFs [44].

mixing of the metal-containing SBUs and mixing of both of the SBUs and organic linkers within the same MOFs backbone that cause heterogeneity within MOFs' order. But recently, SBBs (supramolecular building blocks) instead of SBUs have been applied to construction of MOFs that have more complexity rather than SBUs [28]. These complexities will cause the topology of MOFs to improve than the old structure [29–33]. The supermolecular building layers (SBLs) approach is a unique and full force of method that by Eddaoudi *et al.* in 2011 was presented to create pillared MOFs with much more complexity level [34]. Afterward, MOFs were synthesized with hollow structures and high complexity. The studies showed that the hollow architecture with the different sizes (as micro/nanosized) can have complexity due to the heterogeneity in the morphology or chemical composition. Considering that these compounds have unique properties in the vast variety of applications such as drug delivery, catalysis, and optics, so it is anticipated that the hollow materials are strongly desirable and suitable for basic studies [35–39]. Until now, we have just briefly discussed about MOFs with their properties [40–44] and the effective factors of complexity along with some examples [45], but from the next chapters, the further explanations will be thoroughly described.

2.2 Conclusion

There are many routes to describe complexity in materials as MOFs. Complexity concept means that a series of agents cause MOFs with homogeneous structures to be converted to their heterogeneous counterparts. In this chapter, we gave a brief introduction to the notion of complexity as a concept relevant to structures of MOFs. Understanding complex systems of MOFs is actually a deep issue. There are various levels of complexity in MOFs that can be generated by different approaches. Here, we abstractly introduced the variety of the heterogeneities which produce the complexity in MOFs such as complexity based on building blocks (mixed/polytopic/multiheterotopic ligands, mixed metal nodes, and multiple SBUs), frameworks (chiral frameworks, structural defects, and heterogeneous pores), and mixed MOFs along with some examples. Structural chemists and crystallographers recognize that crystals have a higher level of complexity than molecules. However, we think all kinds of materials can have complex structures and behaviors, too.

References

1. Zarekarizi, F. and Morsali, A., Ultrasonic-assisted synthesis of nano-sized metal-organic framework; a simple method to explore selective and fast Congo Red adsorption. *Ultrason. Sonochem.*, *69*, 105246, 2020.
2. Laha, S., Chakraborty, A., Maji, T.K., Synergistic Role of Microwave and Perturbation toward Synthesis of Hierarchical Porous MOFs with Tunable Porosity. *Inorg. Chem.*, *59*, 6, 3775–3782, 2020.
3. Kitagawa, S., Kitaura, R., Noro, S.I, Functional porous coordination polymers. *Angew. Chem. Int. Ed.*, *43*, 18, 2334–2375, 2004.
4. Kondo, M., Furukawa, S., Hirai, K., Kitagawa, S., Coordinatively immobilized monolayers on porous coordination polymer crystals. *Angew. Chem. Int. Ed.*, *49*, 31, 5327–5330, 2010.
5. Yanai, N. and Granick, S., Directional Self-Assembly of a Colloidal Metal–Organic Framework. *Angew. Chem. Int. Ed.*, *51*, 23, 5638–5641, 2012.
6. Song, X., Jeong, S., Kim, D., Lah, M.S., Transmetalations in two metal–organic frameworks with different framework flexibilities: Kinetics and core–shell heterostructure. *CrystEngComm*, *14*, 18, 5753–5756, 2012.
7. Song, X., Kim, T.K., Kim, H., Kim, D., Jeong, S., Moon, H.R., Lah, M.S., Post-synthetic modifications of framework metal ions in isostructural metal–organic frameworks: Core–shell heterostructures via selective transmetalations. *Chem. Mater.*, *24*, 15, 3065–3073, 2012.
8. Qin, J.-S., Du, D.-Y., Li, M., Lian, X.-Z., Dong, L.-Z., Bosch, M., Su, Z.-M., Zhang, Q., Li, S.-L., Lan, Y.-Q., Yuan, S., Derivation and Decoration of Nets with Trigonal-Prismatic Nodes: A Unique Route to Reticular Synthesis of Metal–Organic Frameworks. *J. Am. Chem. Soc.*, *138*, 16, 5299–5307, 2016.
9. Inge, A.K., Köppen, M., Su, J., Feyand, M., Xu, H., Zou, X., O'Keeffe, M., Stock, N., Unprecedented topological complexity in a metal–organic framework constructed from simple building units. *J. Am. Chem. Soc.*, *138*, 6, 1970–1976, 2016.

10. Alezi, D., Spanopoulos, I., Tsangarakis, C., Shkurenko, A., Adil, K., Belmabkhout, Y., O' Keeffe, M., Eddaoudi, M., Trikalitis, P.N., Reticular chemistry at its best: directed assembly of hexagonal building units into the awaited metal-organic framework with the intricate polybenzene topology, pbz-MOF. *J. Am. Chem. Soc.*, *138*, 39, 12767–12770, 2016.
11. Furukawa, H., Müller, U., Yaghi, O.M., "Heterogeneity within order" in metal–organic frameworks. *Angew. Chem. Int. Ed.*, *54*, 11, 3417–3430, 2015.
12. Yu, D., Shao, Q., Song, Q., Cui, J., Zhang, Y., Wu, B., Ge, L., Wang, Y., Zhang, Y., Qin, Y., Vajtai, R., A solvent-assisted ligand exchange approach enables metal-organic frameworks with diverse and complex architectures. *Nat. Commun.*, *11*, 1, 1–10, 2020.
13. Planes, O.M., Schouwink, P.A., Bila, J.L., Fadaei-Tirani, F., Scopelliti, R., Severin, K., Incorporation of Clathrochelate-Based Metalloligands in Metal–Organic Frameworks by Solvent-Assisted Ligand Exchange. *Cryst. Growth Des.*, *20*, 3, 1394–1399, 2020.
14. Karagiaridi, O., Bury, W., Mondloch, J.E., Hupp, J.T., Farha, O.K., Solvent-assisted linker exchange: an alternative to the de novo synthesis of unattainable metal–organic frameworks. *Angew. Chem. Int. Ed.*, *53*, 18, 4530–4540, 2014.
15. Kalaj, M., Prosser, K.E., Cohen, S.M., Room temperature aqueous synthesis of UiO-66 derivatives via postsynthetic exchange. *Dalton Trans.*, *49*, 26, 8841–8845, 2020.
16. Deria, P., Mondloch, J.E., Karagiaridi, O., Bury, W., Hupp, J.T., Farha, O.K., Beyond post-synthesis modification: evolution of metal–organic frameworks via building block replacement. *Chem. Soc. Rev.*, *43*, 16, 5896–5912, 2014.
17. Shi, D., Cui, C.-J., Hu, M., Ren, A.-H., Song, L.-B., Liu, C.-S., Du, M., A microporous mixed-metal (Na/Cu) mixed-ligand (flexible/rigid) metal–organic framework for photocatalytic H_2 generation. *J. Mater. Chem. C*, *7*, 33, 10211–10217, 2019.
18. Morris, R.E. and Bu, X., Induction of chiral porous solids containing only achiral building blocks. *Nat. Chem.*, *2*, 5, 353, 2010.
19. Kanoo, P., Haldar, R., Cyriac, S.T., Maji, T.K., Coordination driven axial chirality in a microporous solid assembled from an achiral linker via *in situ* C–N coupling. *Chem. Commun.*, *47*, 39, 11038–11040, 2011.
20. Vermoortele, F., Bueken, B., Le Bars, G. l., Van de Voorde, B., Vandichel, M., Houthoofd, K., Vimont, A., Daturi, M., Waroquier, M., Van Speybroeck, V., Kirschhock, C., Synthesis modulation as a tool

to increase the catalytic activity of metal–organic frameworks: the unique case of UiO-66 (Zr). *J. Am. Chem. Soc.*, *135*, 31, 11465–11468, 2013.
21. Xiang, W., Ren, J., Chen, S., Shen, C., Chen, Y., Zhang, M., Liu, C.J., The metal–organic framework UiO-66 with missing-linker defects: A highly active catalyst for carbon dioxide cycloaddition. *Appl. Energy*, *277*, 115560, 2020.
22. Wu, H., Chua, Y.S., Krungleviciute, V., Tyagi, M., Chen, P., Yildirim, T., Zhou, W., Unusual and highly tunable missing-linker defects in zirconium metal–organic framework UiO-66 and their important effects on gas adsorption. *J. Am. Chem. Soc.*, *135*, 28, 10525–10532, 2013.
23. Xue, Z., Liu, K., Liu, Q., Li, Y., Li, M., Su, C.Y., Ogiwara, N., Kobayashi, H., Kitagawa, H., Liu, M., Li, G., Missing-linker metal-organic frameworks for oxygen evolution reaction. *Nat. Commun.*, *10*, 1, 1–8, 2019.
24. Fang, Z., Dürholt, J.P., Kauer, M., Zhang, W., Lochenie, C., Jee, B., Albada, B., Metzler-Nolte, N., Pöppl, A., Weber, B., Muhler, M., Structural complexity in metal–organic frameworks: Simultaneous modification of open metal sites and hierarchical porosity by systematic doping with defective linkers. *J. Am. Chem. Soc.*, *136*, 27, 9627–9636, 2014.
25. Razavi, S.A.A. and Morsali, A., Linker functionalized metal-organic frameworks. *Coord. Chem. Rev.*, *399*, 213023, 2019.
26. Pang, J., Yuan, S., Qin, J., Wu, M., Lollar, C.T., Li, J., Huang, N., Li, B., Zhang, P., Zhou, H.-C., Enhancing Pore-Environment Complexity Using a Trapezoidal Linker: Toward Stepwise Assembly of Multivariate Quinary Metal–Organic Frameworks. *J. Am. Chem. Soc.*, *140*, 39, 12328–12332, 2018.
27. Liu, C., Luo, T.-Y., Feura, E.S., Zhang, C., Rosi, N.L., Orthogonal Ternary Functionalization of a Mesoporous Metal–Organic Framework via Sequential Postsynthetic Ligand Exchange. *J. Am. Chem. Soc.*, *137*, 33, 10508–10511, 2015.
28. Xuan, W., Zhu, C., Liu, Y., Cui, Y., Mesoporous metal–organic framework materials. *Chem. Soc. Rev.*, *41*, 5, 1677–1695, 2012.
29. Yan, Y., Telepeni, I., Yang, S., Lin, X., Kockelmann, W., Dailly, A., Blake, A.J., Lewis, W., Walker, G.S., Allan, D.R., Barnett, S.A., Metal–organic polyhedral frameworks: high H_2 adsorption capacities and neutron powder diffraction studies. *J. Am. Chem. Soc.*, *132*, 12, 4092–4094, 2010.
30. Yuan, D., Zhao, D., Sun, D., Zhou, H.C., An Isoreticular Series of Metal–Organic Frameworks with Dendritic Hexacarboxylate Ligands

and Exceptionally High Gas-Uptake Capacity. *Angew. Chem. Int. Ed.*, *49*, 31, 5357–5361, 2010.
31. (a) Moulton, B., Lu, J., Mondal, A., Zaworotko, M., Nanoballs: nanoscale faceted polyhedra with large windows and cavities. *Chem. Commun.*, 9, 863–864, 2001; (b) Eddaoudi, M., Kim, J., Wachter, J., Chae, H., O'keeffe, M., Yaghi, O.M., Porous metal–organic polyhedra: 25 Å cuboctahedron constructed from 12 $Cu_2(CO_2)$ 4 paddle-wheel building blocks. *J. Am. Chem. Soc.*, *123*, 18, 4368–4369, 2001.
32. Delgado-Friedrichs, O. and O'Keeffe, M., Three-periodic tilings and nets: face-transitive tilings and edge-transitive nets revisited. *Acta Crystallogr. Sect. A: Found. Crystallogr.*, *63*, 4, 344–347, 2007.
33. O'Keeffe, M., Peskov, M.A., Ramsden, S.J., Yaghi, O.M., The reticular chemistry structure resource (RCSR) database of, and symbols for, crystal nets. *Acc. Chem. Res.*, *41*, 12, 1782–1789, 2008.
34. Eubank, J.F., Mouttaki, H., Cairns, A.J., Belmabkhout, Y., Wojtas, L., Luebke, R., Alkordi, M., Eddaoudi, M., The quest for modular nanocages: tbo-MOF as an archetype for mutual substitution, functionalization, and expansion of quadrangular pillar building blocks. *J. Am. Chem. Soc.*, *133*, 36, 14204–14207, 2011.
35. Fuji, M., Han, Y.S., Takai, C.J.K.P., Journal, P., Synthesis and applications of hollow particles. *KONA Powder Part. J.*, *30*, 47–68, 2013.
36. Zhang, L., Wu, H.B., Lou, X.W., Metal–organic-frameworks-derived general formation of hollow structures with high complexity. *J. Am. Chem. Soc.*, *135*, 29, 10664–10672, 2013.
37. Patel, U., Parmar, B., Patel, P., Dadhania, A., Suresh, E., The synthesis and characterization of Zn (ii)/Cd (ii) based MOFs by a mixed ligand strategy: a Zn (ii) MOF as a dual functional material for reversible dye adsorption and as a heterogeneous catalyst for the Biginelli reaction. *Mater. Chem. Front.*, *5*, 1, 304–314, 2020.
38. Bhadu, G.R., Parmar, B., Patel, P., Paul, A., Chaudhari, J.C., Srivastava, D.N., Suresh, E., Co@ N-doped carbon nanomaterial derived by simple pyrolysis of mixed-ligand MOF as an active and stable oxygen evolution electrocatalyst. *Appl. Surf. Sci.*, *529*, 147081, 2020.
39. Razavi, S.A.A., Berijani, K., Morsali, A., Hybrid nanomaterials for asymmetric purposes: green enantioselective C–C bond formation by chiralization and multi-functionalization approaches. *Catal. Sci. Technol.*, *10*, 24, 8240–8253, 2020.
40. Robison, L., Gong, X., Evans, A.M., Son, F.A., Wang, X., Redfern, L.R., Wasson, M.C., Syed, Z.H., Chen, Z., Idrees, K.B., Islamoglu, T., Transient catenation in a zirconium-based metal–organic framework

and its effect on mechanical stability and sorption properties. *J. Am. Chem. Soc., 143*, 3, 1503–1512, 2021.
41. Du, J., Li, F., Sun, L., Metal-organic frameworks and their derivatives as electrocatalysts for the oxygen evolution reaction. *Chem. Soc. Rev., 50*, 4, 2663–2695, 2021.
42. Berijani, K. and Morsali, A., Construction of an asymmetric porphyrinic zirconium metal-organic framework through ionic postchiral modification. *Inorg. Chem., 60*, 1, 206–218, 2020.
43. Chakraborty, G., Park, I.H., Medishetty, R., Vittal, J.J., Two-dimensional metal-organic framework materials: synthesis, structures, properties and applications. *Chem. Rev., 121*, 7, 3751–3891, 2021.
44. Fang, Z., Dürholt, J.P., Kauer, M., Zhang, W., Lochenie, C., Jee, B., Albada, B., Metzler-Nolte, N., Pöppl, A., Weber, B. and Muhler, M., Structural complexity in metal-organic frameworks: Simultaneous modification of open metal sites and hierarchical porosity by systematic doping with defective linkers. *J. Am. Chem. Soc.*, 136, 27, 9627-9636, 2014.
45. Mobus, G.E. and Kalton, M.C., Principles of systems science: 755, Springer-Verlag, New York, ISBN: 13 9781493919208, 2015.

3

Complexity Based on Ligand—Part 1

Abstract
The large majority of the synthesized MOFs have relatively 'simple' building units. But, recently, complex MOFs have been constructed through different strategies such as complex building units or mixed components. The progress of synthesis methods toward MOFs with more complex is very important. In this field, there are challenges such as how inducing heterogeneity into structure, their characterization methods and maintenance of framework stability. One of effective methods to produce complexity in MOFs is using mixed ligand. The use of mixtures of organic-building block linkers can lead to chemical complexity and deviation from homogeneity of framework. According to that, the mixed component MOFs and mixed ligand effect on MOF complexity are not extensively studied yet, so in this chapter, we discuss mixed ligand MOFs along with some of examples and how their effect on the complexity generation.

***Keywords*:** Mixed ligand, chemical complexity, backbone heterogeneity, structural tunability

3.1 Mixed Ligand

Many crystal lattices of MOFs were synthesized with the different designing and structural properties by using the various approaches [1–5]. The integration of various building blocks, which property the same structure and connectivity, can lead to the production of MOFs with mixed component (MIXMOFs). The MIXMOF concept can lead to modern materials with complex structures and unusual features. Until now, mixed component MOFs (mixed-metal and mixed-linker MOFs) have not been extensively investigated. We want to

Ali Morsali and Kayhaneh Berijani. Metal-Organic Frameworks with Heterogeneous Structures, (27–56) © 2021 Scrivener Publishing LLC

address the most significant ones in the following and start with mixed-linker MOFs. There are several types of mixed-linker MOFs such as pillared-layer mixed-linker, cage-directed mixed-linker, cluster-based mixed-linker, and structure templated mixed-linker MOFs that herein, some of them will be explained. Commonly, the used organic ligands in the secondary building units of MOFs were one kind of ditopic linker. But, recently, two or more different kinds of ligands are also used in the preparation of MOFs framework. With more investigations about this type of MOFs, it has been shown that the using two or more different types of organic linker is a useful and efficient method to induce heterogeneity to MOFs [6]. These MOFs can be created by using the mixed bidentate linkers, and polytopic and multi-heterotopic ligands that two last concepts will be discussed in the next sections. A wide range of MOFs was synthesized by using the various linkers as organic precursors with different features as length, geometry, a kind of atom donor (like N-, O-), and functional groups in structure. The designable framework of MOF, due to tunable components, can make changing linkers possible that it can cause the heterogeneity in MOF structure so the synthesis of this type of MOFs is interesting.

Based on the used linkers in MOF design, the matrix of MOF can have different vacancies that during the insertion of the various linkers with different properties (like length, geometry, and functional group) to MOF structure, the complicated crystal lattice can be obtained with high complexity along with complex pores [7]. Several models in this kind of complexity have been reported. At first, in 2001, two types of organic linkers were used for synthesizing of Cu-based MOF. The results showed that the storage capacity of the synthesized coordination polymer has increased [8]. Then, several MOFs with various mixed ligands have been prepared until now that some of them have been illustrated in Table 3.1. The investigations show that the type of linker can influence on the properties and applications of MOFs. A series of photocatalysts based on MOF was synthesized by mixed ligands. One kind of Re-MOF was synthesized with SBUs from $Zr_6O_4(OH)_4(-CO_2)_{12}$ clusters and three different ditopic linkers (H_2BPDC, H_2ReTC, and $H_2BPDC\text{-}(NH_2)_2$) (Figure 3.1) [9]. This MOF with multiple functional ligands as a photo active catalyst demonstrated high selectivity and activity in

Table 3.1 Examples of mixed ligands metal-organic frameworks.

Entry	Mixed-ligand metal-organic frameworks	The used mixed ligands	Ref.
1	$\{[Zn(azbpy)(HO\text{-}1,3\text{-}bdc)(H_2O)].(azbpy)\}_n$	azbpy: 4,4′-azobispyridine HO-1,3-bdc: 5-hydroxy isophthalate	[27]
2	$\{[Zn_2(bdc)_2(4\text{-}bpdh)].C_2H_5OH.2H_2O\}_n$	bdc^{2-}: dianion of 1,4 benzene dicarboxylic acid 4-bpdh: 2,5-bis(4-pyridyl)-3,4-diaza-2,4-hexadiene	[28]
3	$[Zn(bdc)(4\text{-}bpdh)]_n$		
4	$\{[Zn_2(bdc)_2(4\text{-}bpdh)_2].(4\text{-}bpdh)\}_n$		
5	$\{[Zn(bdc)(4\text{-}bpdh)].C_2H_5OH\}_n$		
6	Mixed ligands Cu-metal organic frameworks (MOFs-2 and MOFs-3)	BTC: 1,3,5-benzene tricarboxylate IPA: isophthalic acid	[29]
7	Re-MOF-$(NH_2)^1$	Re: $ReI(CO)_3(BPYDC)(Cl)$ BPYDC: 2,2′-bipyridine-5,5′-dicarboxylate BPDC-$(NH_2)_2$: 2,2′-diaminobiphenyl-4,4′-dicarboxylate BPDC: 4,4′-biphenyldicarboxylate	[9]
8	$[Cd_3(bdc)(HCOO)_2(tipo)_2(H_2O)_2]\cdot 2NO_3\cdot 6DMF$	bdc^{2-}: phenyl-1,4-dicarboxylate tipo: tris[4-(1H-imidazol-1-yl) phenyl] phosphine oxide $bpdc^{2-}$: biphenyl-4,4'-dicarboxylate	[30]
9	$[Zn_8(OH)_4(bpdc)_6(tipo)_4]\cdot 16DMF$		

(Continued)

Table 3.1 Examples of mixed ligands metal-organic frameworks. (*Continued*)

Entry	Mixed-ligand metal-organic frameworks	The used mixed ligands	Ref.
10	$[Ni(L)_{0.5}(4,4'\text{-bipy})_{0.5}(H_2O)_2]_n$	H_4L: 1,3-di(3′,5′-dicarboxylphenyl) benzene 4,4′-bipy: 4,4′-bipyridine bpp: 1,3-di(4-pyridyl)propane	[31]
11	$[Zn_2(L)(bpp)_2]_n$		
12	$[(CH_3)_2NH_2]_4[Zn_4O]_4[Zn(TCPP)]_5[BTB]^2_{8/3}$	BTB: 1,3,5-benzene(tris)benzoate TCPP: tetrakis(4-carboxyphenyl) porphyrin TBTB: 4,4′,4″-(2,4,6-trimethyl benzene-1,3,5-triyl) tribenzoate	[32]
13	$[Zr_6(\mu_3\text{-}O)_4(\mu_3\text{-}OH)_4][TCPP][TBTB]^3_{8/3}$		
14	$Zn_2(NDC)_2(DPNI)$	NDC: 2,6-naphthalenedicarboxylate DPNI: N,N′-di-(4-pyridyl)-1,4,5,8-naphthalene tetracarboxydiimide	[33]
15	$[Cu(2,4\text{-pydca})(imidazole)_2] \cdot 2H_2O$	2,4-pydca: 2,4-pyridinedicarboxylate	[34]
16	$[Co_2(ma)(ina)]_n \cdot 2nH_2O$	ma: malate ina: isonicotinate pico: 3-hydroxypicolinate suc: succinate H_2hypa: hydroxy-phenyl-acetic acid 4,4'-bpy: 4,4′-bipyridine pzdc: 2,3-pyrazinedicarboxylate	[5]
17	$[Co_3(ina)_2(pico)_2(H_2O)_2]_n$		
18	$[Mn_3(suc)_2(ina)_2]_n$		
19	$[Co_2(hypa)_2(4,4'\text{-bpy})]$		
20	$\{[Cd_2(pzdc)_2(L_1)(H_2O)_2] \cdot 5H_2O \cdot CH_3CH_2OH\}_n$		

(*Continued*)

Table 3.1 Examples of mixed ligands metal-organic frameworks. (*Continued*)

Entry	Mixed-ligand metal-organic frameworks	The used mixed ligands	Ref.
21	$[Fe(pydc)(4,4'\text{-}bpy)]\cdot H_2O$	L_1: 2,5-bis(2-hydroxyethoxy)-1,4-bis(4-pyridyl) benzene pydc: 2,5-pyridinedicarboxylate $5\text{-}NH_2\text{-}bdc$: 5-aminoisophthalate azpy: 4,4′-azopyridine pyrdc: pyridine-2,3-dicarboxylate bpndc: benzophenone-4,4′-dicarboxylate bdc:1,4-benzenedicarboxylate L-lac: L-lactate	
22	$[Co(5\text{-}NH_2\text{-}bdc)(4,4'\text{-}bpy)0.5(H_2O)]3\cdot 2H_2O$		
23	$\{[Cd_4(azpy)_2(pyrdc)4(H_2O)_2]\cdot 9H_2O\}_n$		
24	$\{[Cd(bpndc)(4,4'\text{-}bpy)](DMF)(H_2O)\}_n$		
25	$[Zn_2(bdc)(L\text{-}lac)(DMF)]\cdot(DMF)$		
26	Anionic MOF $[Zn_8(ad)_4(bdc)_6O\cdot 2Me_2NH_2\cdot 8DMF\cdot 11H_2O]$	ad: adeninate	[35]
27	MOF-500 $[NH2(CH3)_2^+]_8[(Fe_3O)4(SO4)_{12(}bdc)_{6(}bpe)6^-]_8$ .13H2O.8DMF	bpe: cis-1,2-bis-4-pyridylethane	[36]
28	$[Zn_2(2,6\text{-}ndc)_2(2\text{-}Pn)]$	2-Pn: 2-Pn: N,N′-bis(pyridin-2-yl) benzene-1,4-diamine H_2cca: 4-carboxycinnamic acid	[37]
29	$[Zn_2(cca)_2(2\text{-}Pn)]$		
30	$[Zn_2(fma)(trz)_2]$	H_2fma: fumaric acid Htrz : 1,2,4-triazole	[38]

(*Continued*)

Table 3.1 Examples of mixed ligands metal-organic frameworks. (*Continued*)

Entry	Mixed-ligand metal-organic frameworks	The used mixed ligands	Ref.
31	MOF-5-NH_2	ABDC: 2-amino-1,4-benzene dicarboxylic acid	[39]
32	$[Co_4(pico)_4(4,4'\text{-bpy})_3(H_2O)_2]_n \cdot 2nH_2O$	pico^{2-}: 3-hydroxypicolinate	
33	PCN-609-BDC-BPDC	TPDC: 2',5'-dimethylterphenyl-4,4"-dicarboxylate	[40]
34	PCN-609-BDC-TPDC		
35	PCN-609-BPDC-TPDC		
36	PCN-609-BDC-BPDC-TPDC		

[1] $Zr_6O_4(OH)_4(-CO_2)12$ secondary building units.
[2] PCN-137.
[3] PCN-138.

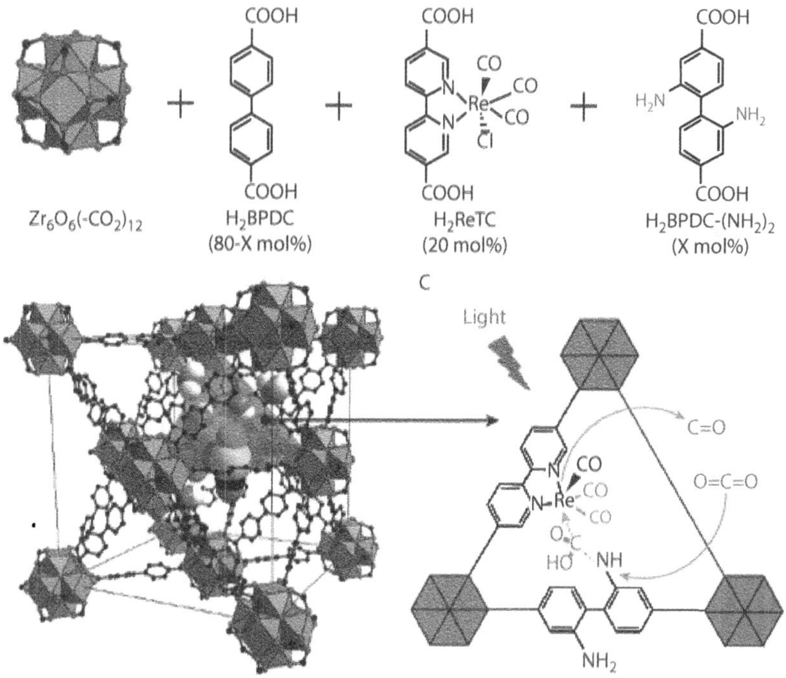

Figure 3.1 A photocatalyst MOF with three different ditopic linkers [9].

conversion of CO_2 to valuable CO. In the construction of this MOF, ligand and metal components have been prepared separately. In a glass vial, 0.04 mmol $ZrCl_4$ was dissolved in the presence of 5 ml of DMF and 0.5 ml of acetic acid. In the synthesis of ligand, 0.008 mmol rhenium ligands, H_2-BPDC-$(NH_2)_2$ (from 0 to 80 mole percent of H_2ReTC), biphenyl-4,4′-dicarboxylic acid, and 5 ml of N,N'-dimethylformamide were used. In the following, both ligand and metal solutions were added into vial that was placed in oven with certain conditions (85°C/12 h).

Finally, the obtained MOF was washed with determined solvents and dried by using a vacuum drying oven. In the mentioned photocatalytic conversion, the optimized amount of -NH_2 functional groups was 33 mol%. These findings indicated that functional groups affected on the bond lengths of Re-CO in ReTC and heterogeneity enhanced photocatalytic activity of MOF. In the synthesis of mixed ligand MOFs, the metallo-linkers should not be forgotten as shown in Figure 3.1 (H_2ReTC). Structural analysis of the generated samples was performed by several techniques such as PXRD patterns, SEM images, N2 adsorption isotherms, and IR spectra (Figure 3.2). As mentioned, percent incorporation of BPDC-$(HN_2)_2$ in framework was also investigated.

Metallosen ligands have formerly been employed in construction of MOFs with two different linkers. For instance, [$Zn_2(bpdc)_2$ dipyridyl Mn(salen)] had a 2D structure with suitable pores and it was used in catalytic reactions as a heterogeneous porous catalyst [10]. According to the importance of drug delivery, the ability of some of mixed ligands MOFs in this application was investigated, too [11]. Undoubtedly, the controlling porosity and pore size due to the organic linkers in the preparation of the hierarchical porous MOFs

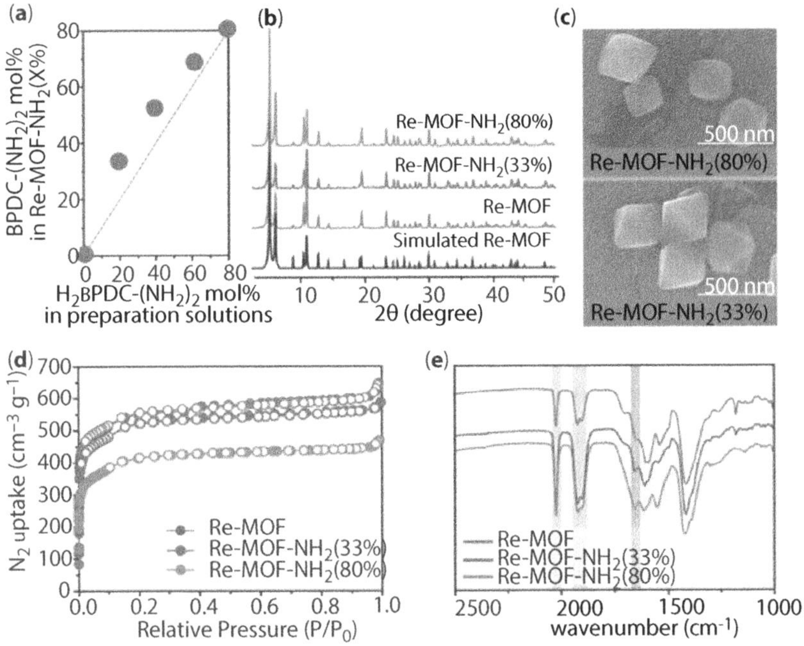

Figure 3.2 Structural analyses of a MOF-based photocatalyst with three different ditopic linkers. (a) Percent of BPDC-$(HN_2)_2$ incorporation in ReMOF-NH_2 (X%). (b) PXRD patterns: Re-MOF, Re-MOF-NH_2 (33%), Re-MOF-NH_2 (80%) and Re-MOF simulated pattern. (c) SEM images: Re-MOF-NH_2 (33%) and Re-MOF-NH_2 (80%). (d) N_2 adsorption isotherms: Re-MOF, Re-MOF-NH_2 (33%) and Re-MOF-NH_2 (80%). (e) IR spectra: Re-MOF, Re-MOF-NH_2 (33%) and Re-MOF-NH_2 (80%) [9].

can affect both adsorption of drug and its delivery. Even, this kind of chemical complexity in MOFs structure can generate conditions for different uses as gas sorption [12] and separation of gases mixture [13]. For example, Costantino reported isoreticular mixed-linker MOFs based on using phosphonate. Its synthesis was performed by using heterocyclic N-donor co-ligands. In the construction of these kinds of MOFs, copper(II) acetate, H_8L^1 and H_8L^2 along with N-donor ligands bipy and etbipy were employed (Scheme 3.1).

These layered isoreticular MOFs were generated under mild synthetic conditions (temperature: 80°C/solvent: H_2O). In this synthetic method, L(s) are employed as chelating agents owing to restraining one propagation direction of the considered spacers linked to the phosphonic moieties. These materials have different coordination environments around composite building units (CBUs): Octahedral Cu1 and square pyramidal Cu2 (Figure 3.3).

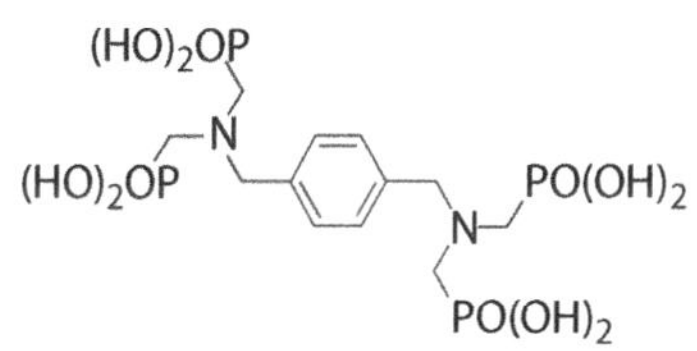

H_8L^1

N,N,N,′N,′- tetrakis (phosphonomethyl) –α,α,′–p–xytylenediamine

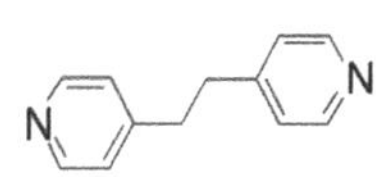

etbipy

1,2–bis (4–pyridyl) ethane

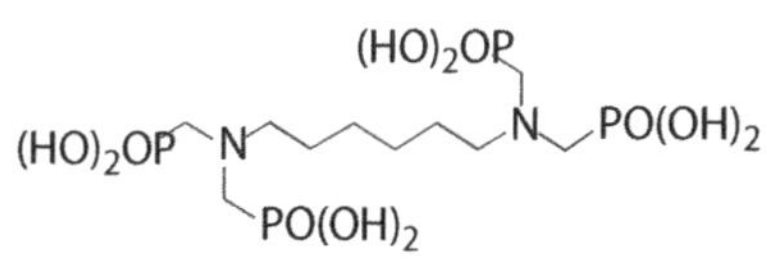

H_8L^2

N,N,N,′N,′- tetrakis (phosphonomethyl) hexamethylenediamine

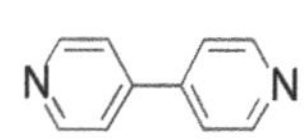

bipy

4,4,′–bipyridine

Scheme 3.1 Structure of the used ligands.

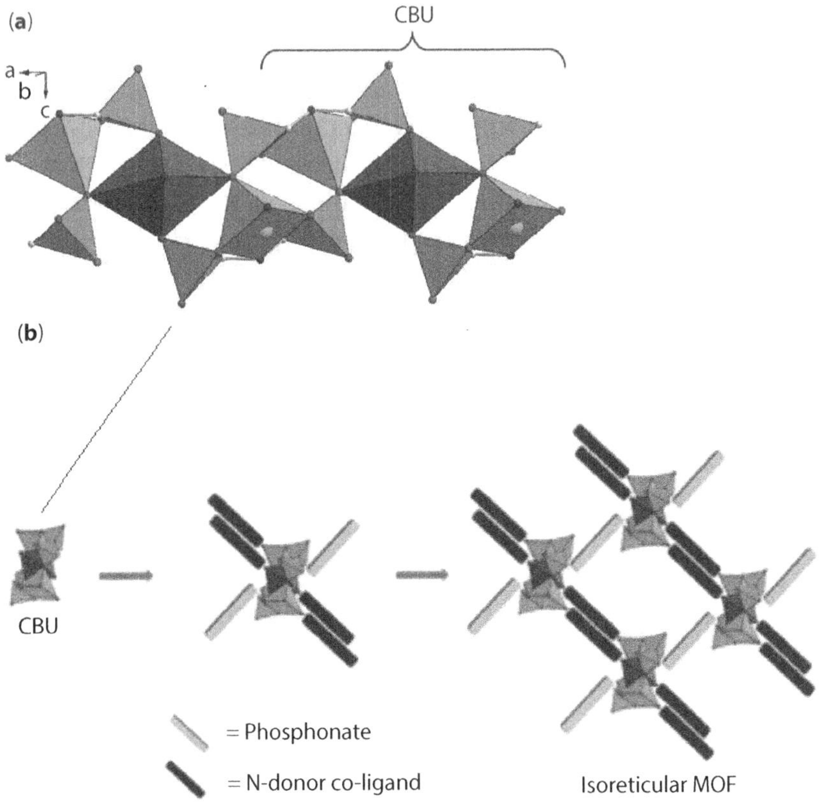

Figure 3.3 (a) Copper-phosphonate CBU polyhedral. Octahedral Cu1 and square pyramidal Cu2: dark blue and light blue, respectively. PO_3C tetrahedral: green. (b) The original structure of the considered isoreticular MOF [12].

According to structural properties of these materials, preliminary gas sorption investigations were made on them, despite that they were non-porous but they can bind to CO_2 at mild conditions. The various behaviors of these isostructure MOFs propose that the used organic linkers play an essential role in their gas sorption.

One of the main drawbacks of MOFs with carboxylate linkers is low resistance to hydrolysis but these unprecedented isoreticular MOFs as water stable frameworks with their structural complexity show uptake capacity and selectivity to CO_2 and N2, respectively. Interestingly, the interpenetrated networks in MOFs can have mixed ligands (mixed ligand interpenetrated MOFs) and they can be candidates as selective adsorbents [13]. Among of metal sites (metal ions or clusters) in the creation of mixed linker MOFs, Zr_6 clusters have been attracted due to the high stability and structural tunability [14–20].

Zhou and coworkers reported the construction of one type of Zr-based MOF with an acidic modulator and two different linkers that are similar from the point of the lengths and connectivity. Its framework can be only supported with one kind of linker and while the other linker was decomposed and removed. During this occurrence, porosity and pore size of MOF increased (Figure 3.4) [21]. The utilized synthesis method in this work is "linker labilization" that leads to the production of mesopores due to crystal defects *via* pro-labile linker splitting and then linker fragments removal by using acid treatment. This process causes that diffusion and adsorption of the considered guest molecules by MOFs happen, easily. In fact, this method has been inspired from the other methods such as linker installation and solvent-assisted ligand incorporation. Figure 3.4 shows the coordinately unsaturated Zr clusters and incorporation of linear linkers or terminal ligands on them. In this process, terminal -OH/H_2O can be replaced by carboxylates (Figure 3.4a, top). In Figure 3.4 (b and c, top), it has been shown that pro-labile linkers are split into two removable monocarboxylates. To study linker exchange, the several methods have been used, one of them is UV-vis spectroscopy that its results have been reported in Figure 3.4 (bottom). However, other techniques analysis were also employed to investigate produced sample like gas sorption measurements, PXRD, SCXRD, NMR, ICP-MS, TGA, IR, TEM, and SEM.

The obtained framework by using this synthesis route along with mixed pores (micro and meso) can enhance activity of immobilized species in framework as enzymes (Figure 3.5). Also, they investigated efficiency of this kind of MOFs in catalysis. Their selected MOF was CYCU-3 that has AZDC linker. Their reason was due to hetero-channels in structure: hexagonal and triangular channels. The large hexagonal channels are suitable for encapsulation of enzyme and substrate diffusion can happen through triangular channels, simply (Figure 3.5, bottom). MOFs can be prepared with mixed-linker by up to 50% substitution of a linker by the other which is important that with this substituting any changing was not observed in the primitive crystal structure [22]. For the production of this type of heterogeneity, a MOF can be also prepared with both the mixed-valence and mixed-linker. In this phenomenon, one of the linkers can be as a defective linker that leads to the isoreticular derivatives of main MOF with heterostructure.

One kind of mixed-valence $Ru^{II/III}$ MOF based on mixed-linker has been reported that it improved the sorption and chemisorption of carbon monoxide and carbon dioxide. (Figure 3.6). In the preparation of this MOF, a mixed-linker solid-solution method was used for modification of metal sites and production of structural defects.

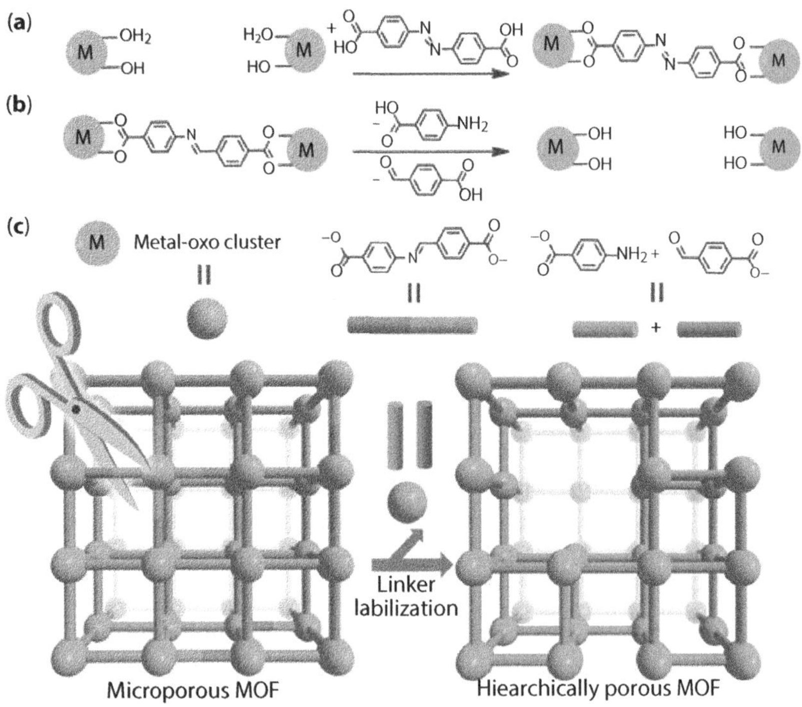

Figure 3.4 Schematic representations of the happened processes in MOF with increased porosity. (a and b) Organic linker installation and linker labilization, respectively. (c) The produced hierarchically porous MOF through linker labilization (top). UV-vis analysis of linker exchange process. (a) Concentration of AZDC in supernatant as a function of incubation time in CBAB solutions with different concentrations. (b) Relationship between CBAB exchanged/CBAB added, exchange ratio, and CBAB molarity. (c) PCN-160 crystals images with various exchange ratios. *(Continued)*

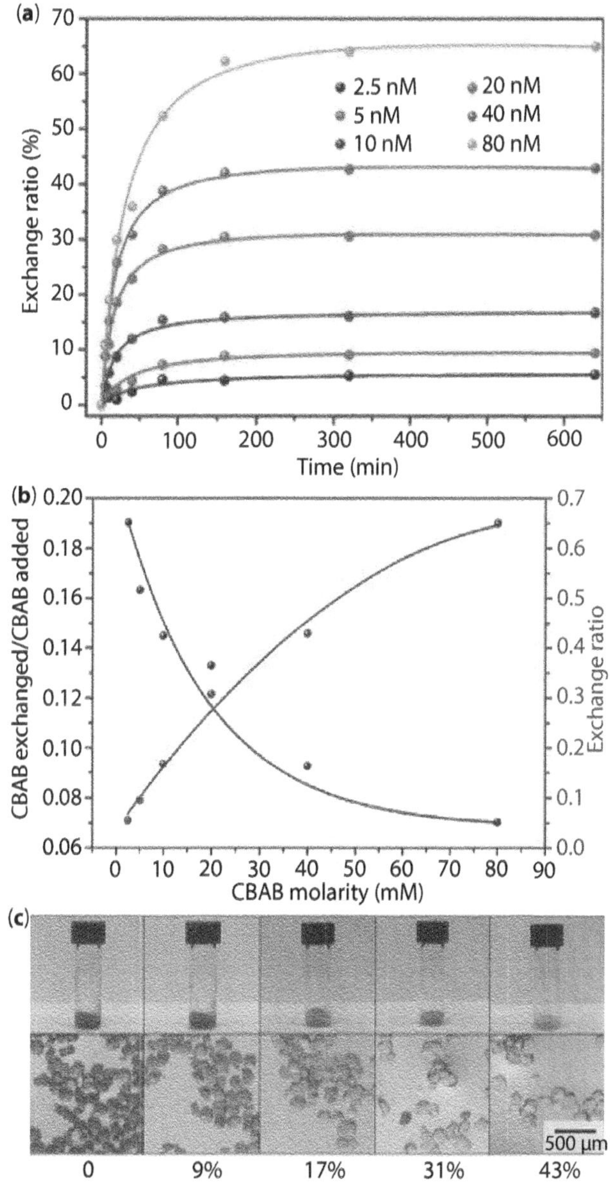

Figure 3.4 (Continued) Schematic representations of the happened processes in MOF with increased porosity. (a and b) Organic linker installation and linker labilization, respectively. (c) The produced hierarchically porous MOF through linker labilization (top). UV-vis analysis of linker exchange process. (a) Concentration of AZDC in supernatant as a function of incubation time in CBAB solutions with different concentrations. (b) Relationship between CBAB exchanged/CBAB added, exchange ratio, and CBAB molarity. (c) PCN-160 crystals images with various exchange ratios [21].

Pyridine-3,5-dicarboxylate (pydc) incorporation which is the same size as BTC but carries lower charge, as a second, defective organic linker has led to Ru-MOF derivatives that are isoreticular along with mixed-linker which show characteristics unlike frameworks that are defect-free.

As well, the formed reduced Ru sites (Ru-hydride) are the positive effects of available defects with the catalytic ability in the hydrogenation

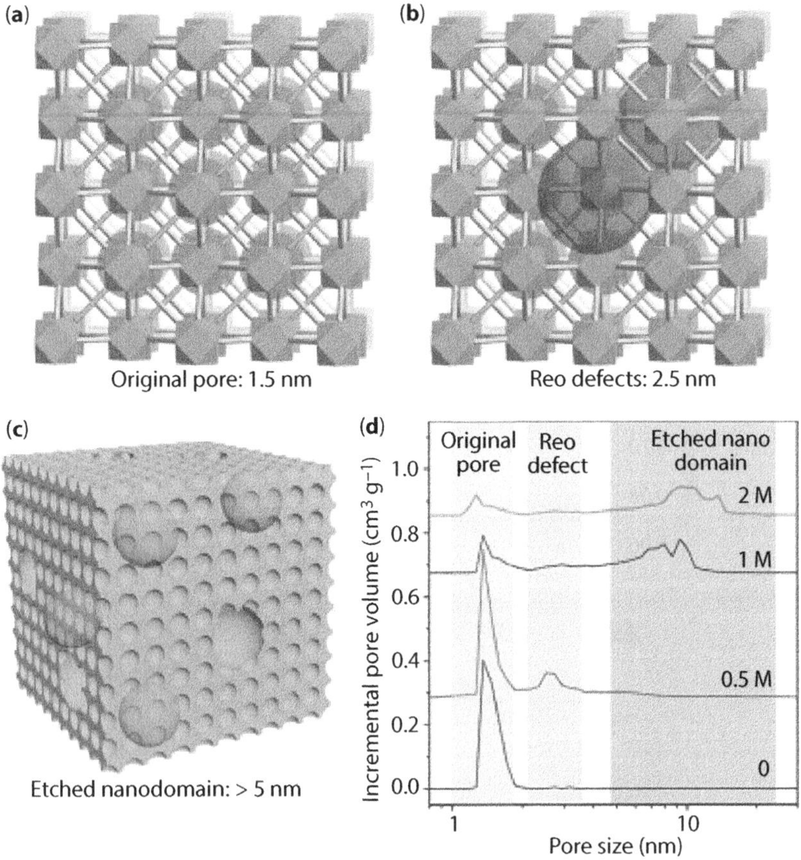

Figure 3.5 Schematic representations of construction mechanism. (a) Formed micropores, (b and c) small and large mesopores, and (d) pores size distribution of the considered MOF (PCN-160-34%) in the presence of acid in different amounts. (Top) (a) Representation of one kind of MOF in the presence of enzyme. (b) relative activity of the considered MOFs in the oxidative reaction of ABTS (2,2′-azino-bis(3-ethylbenzothiazoline-6-sulphonic acid)) and o-PDA (o-phenylenediamine). *(Continued)*

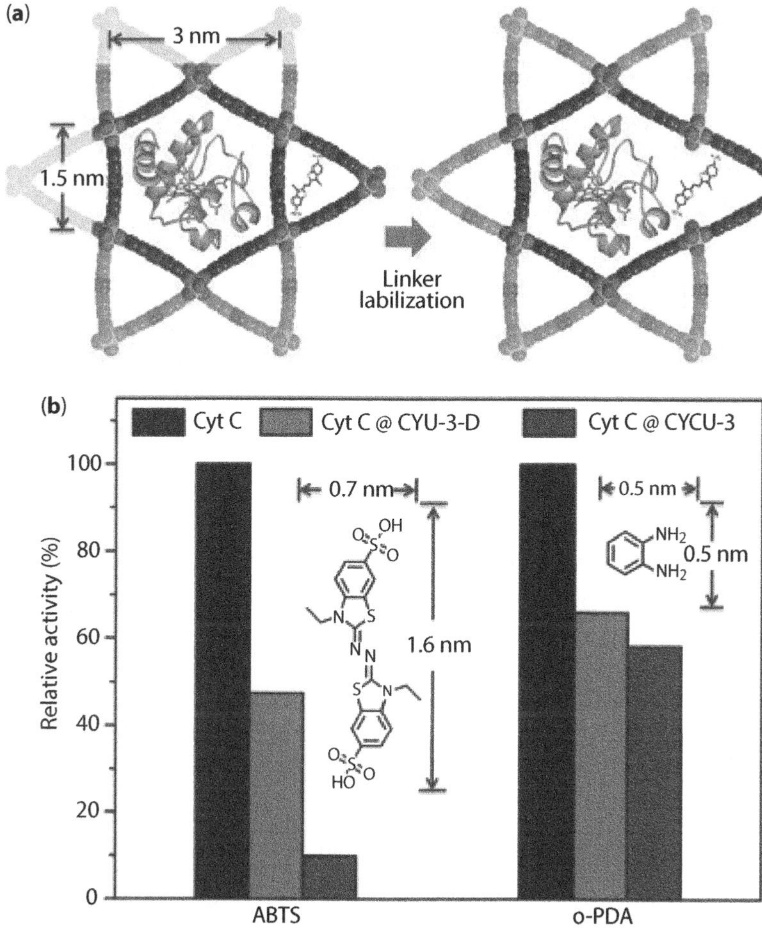

Figure 3.5 (Continued) Schematic representations of construction mechanism. (a) Formed micropores, (b and c) small and large mesopores, and (d) pores size distribution of the considered MOF (PCN-160-34%) in the presence of acid in different amounts. (Top) (a) Representation of one kind of MOF in the presence of enzyme. (b) relative activity of the considered MOFs in the oxidative reaction of ABTS (2,2′-azino-bis(3-ethylbenzothiazoline-6-sulphonic acid)) and o-PDA (o-phenylenediamine) [21].

of olefins, hydride species formation, and increased CO sorption (Figure 3.7) [23]. The other mixed-valence/mixed-ligand MOF as a heterogeneous structure with high porosity was synthesized by Chen and co-workers. $[Ni^{II}_2Ni^{III}(\mu_3\text{–}OH)(pybz)_3(ndc)_{1.5}]\cdot 9.5$ DMA$\cdot$ $8.5H_2O$ has two different metal sites in cluster model, $Ni_3(\mu_3\text{-}OH)$ and

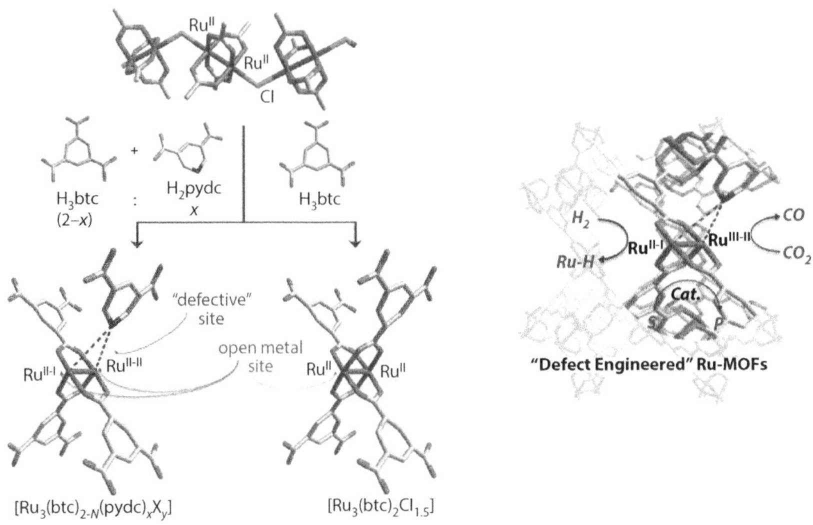

Figure 3.6 One kind of mixed-valence Ru$^{II/III}$ MOF with mixed-linker [23].

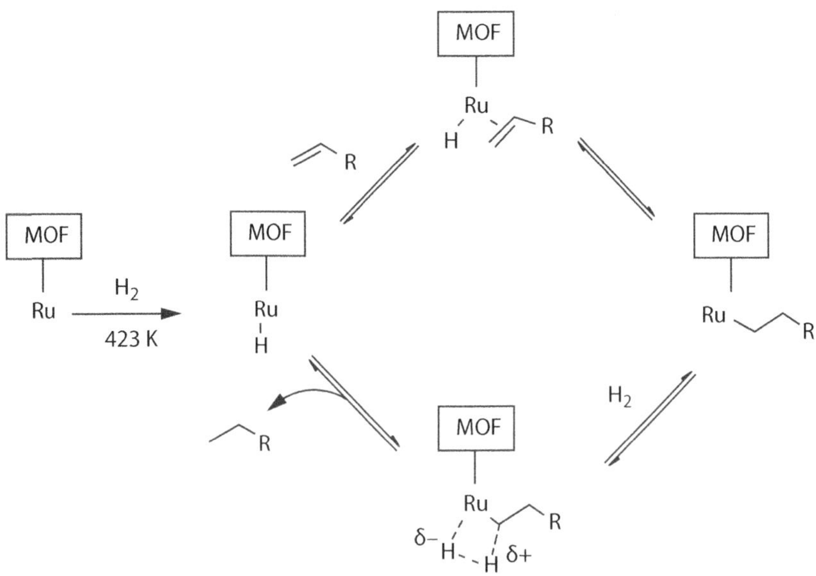

Figure 3.7 Olefin hydrogenation mechanism [23].

$[Ni_3(\mu_3\text{-}OH)(COO)_6]$ and carboxylate ligands in the form of naphthalene-2,6-dicarboxylate and 4-(pyridin-4-yl)benzene carboxylate.

This framework can be heterogeneous structurally having two different kinds of nodes (planer and tricapped trigonal prismatic), two different types of cages trigonal pyramidal and tetrahedral with two different organic ligands. Actually, the matrix of MOF permits two different or more than one kind of organic linkers that can be placed in framework with different shape [24]. The generated complexity in this MOF structure exhibited the improved CO_2 sorption amounts due to the high produced porosity from the structural heterogeneity [25]. MOFs with two different linkers have been remarkably reported in comparison to three and or more types of different linkers. Newly, the preparation of a Zr-based MOF was investigated with three organic linkers and it was interesting that the ordered arrangement was observed in the fragments of framework [26]. Some examples of mixed ligands MOFs in Table 3.1 have been introduced.

Some of these MOFs have been introduced as a programmed-pore MOF (PP-MOF) that creates complexity without losing homogeneity that affects performance of specific applications such as the CO_2 sorption and catalysis and improves these processes [41]. For example, the incorporation of the three different carboxylates as linkers can happen with the production of ultrahigh surface areas and the largest pore volumes that these features can affect MOFs applications [42]. About zeolitic imidazolate frameworks, their modified structures can be also synthesized with using the mixed linkers. In 2012, a new strategy was presented for creation of ZIFs with the various ratios of different linkers to obtain the controlled porosity and functionality. The results showed that the use of these linkers with suitable chemical and thermal stability can influence on the potential for gas separation or catalytic activity of ZIFs. Using bulky imidazole nitrogen atom as a linker is more efficient method for producing the steric complexity (Figure 3.8) [43]. Even, chiral zeolite-like MOFs based on mixed linkers were built, for example, with collection of dipyridyl functionalized chiral Ti (salan) and bpdc (biphenyldicarboxylate) as a linear ditopic linker [44].

A mixed-ligand MOF was synthesized from combination of Ni and two ligands BTC (benzene-1,3,5-tricarboxylate) and BTRE (1,2-bis(1,2,4-triazol-4-yl)-ethane) (Figure 3.9). This MOF is water

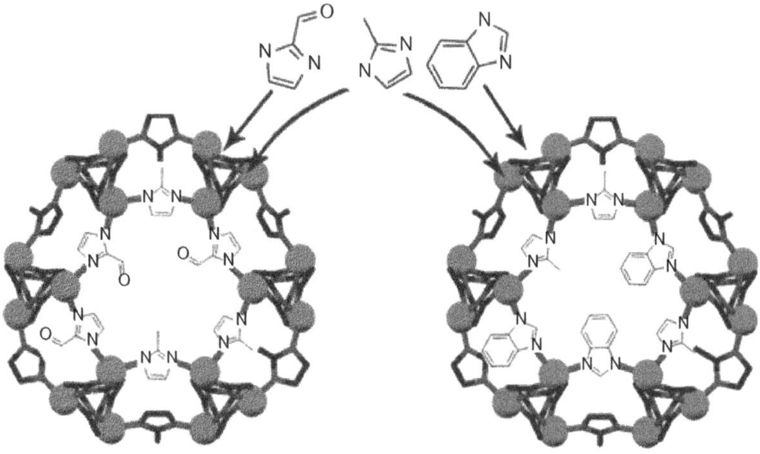

Figure 3.8 Zeolite-like MOFs based on mixed linkers [43].

stable and due to its porous volume has capability of adsorption and desorption of the water to content 30 wt% [45]. MOFs like this MOF, $[Ni_3(\mu_3\text{-}BTC)_2\text{-}(\mu_4\text{-}BTRE)_2(\mu\text{-}H_2O)_2]$, can be employed in mixed-

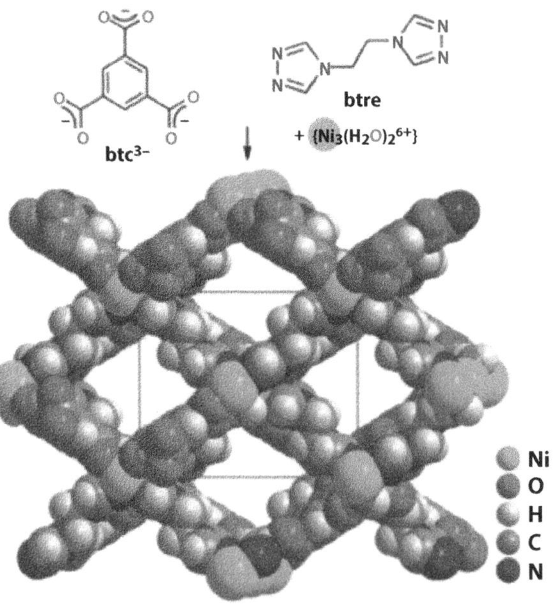

Figure 3.9 A mixed-ligand MOF with two ligands BTC (benzene-1,3,5-tricarboxylate) and BTRE (1,2-bis(1,2,4-triazol-4-yl)-ethane) [45].

matrix membranes (MMMs), gas storage, and separation in chillers and pumps [46–50].

Even, this kind of design, mixed-linker materials, is seen in metal-organic polyhedra (MOPs). For example, Zhou reported a octahedral MOP that was prepared by using 9H-carbazole-3,6-dicarboxylate (CDC) and paddlewheel Cu-cluster. In this work, 4,4′-bipyridine as second ligand was used. The main axial ligand as labile ligand is readily substituted with 4,4′-bipy. In the end, new 3D sample with pcu (pcu-a) topology was formed (Figure 3.10, top) [51]. This design in porphyrin-based MOFs is also seen (Figure 3.10, bottom) [52].

Recently, Suresh *et al.* synthesized Zn(II)- and Cd(II)-based MOFs by mixed ligand strategy with chemical formula $\{[Zn_2(5NO_2\text{-}IP)_2(L)_2](H_2O)\}n$ (ADES-1) (Figure 3.11) and $\{[Cd_2(5NO_2\text{-}IP)_2(L)_2(H_2O)_4](L)(H_2O)(CH_3OH)_6\}n$ (ADES-2) (Figure 3.12) that $5NO_2$-IP and L are 5-Nitroisophthalate and (E)-N′-(pyridin-3-ylmethylene)nicotinohydrazide [53]. Both MOFs were characterized by various methods such as XRD analysis. They were crystallized in the triclinic space group P-1 by maintaining topology of framework.

It is interesting that these kinds of materials can have abilities like dye adsorption. Given that, organic dyes have toxic nature and harmful effects on the environment and human beings, they can be considered as one of the major contaminants of environment specially their presence in industrial waste water. So, adsorption and removal of these compounds that are in general stable, are as environmental challenges. In these processes, chemical stability, pore size tunability, tuning the porosity and topology of MOFs have important role. For example, in this work, aqueous phase dye adsorption studies by ADES-1 was investigated. The chemical structures of the used organic dyes in this work have been also shown in Scheme 3.2.

At first, 100 mg of the considered MOF along with 5×10^{-5} M dye were mixed. This experiment was carried out in the dark and at R.T. (3h). The obtained results showed that apart from nature and charge of dyes, their colour strength reduced so, ADES-1 could adsorb all of them. These changes led to colour change of the first ADES-1 to ADES-1@dye (dye (MV): violet, dye (MB): blue, dye (MO): orange, dye (RhB): pink. Importantly, these kinds of MOFs with

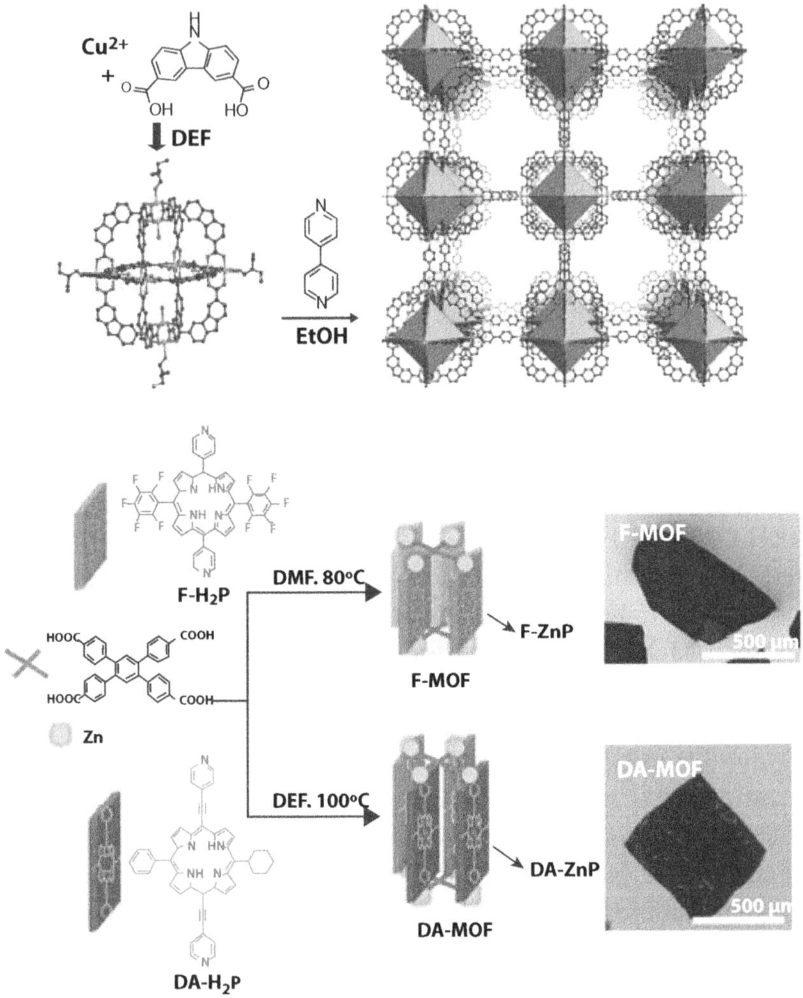

Figure 3.10 Stepwise preparation of $\{[Cu_4(CDC)_4(4,4'\text{-bipy})(H_2O)_2]_3\}_n \cdot xS$. Copper: aqua; oxygen: red; nitrogen: blue; carbon; black. Hydrogen atoms have not been shown (top) [51]. Synthesis of the isostructural porphyrinic MOFs and obtained crystals photographs (bottom) [52].

heterogeneous structure due to mixed ligands can be used in catalysis processes as solid catalyst. Multi-component Biginelli reaction is one of the reactions that can happen by using mixed ligand ADES-1 (Scheme 3.3).

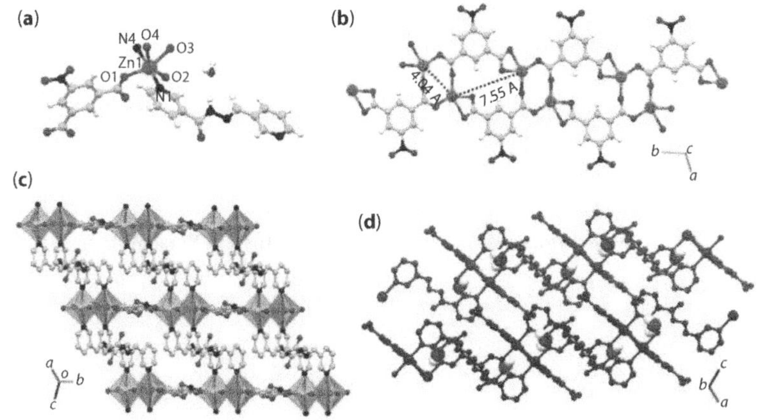

Figure 3.11 ADES-1: (a) coordination environment around Zn(II); (b) and (c) $[Zn_2(COO)_2]$ SBU in 1D metal-carboxylate chain and double lined 2D network, respectively; (d) ABAB manner interlayer with lattice H_2O [53].

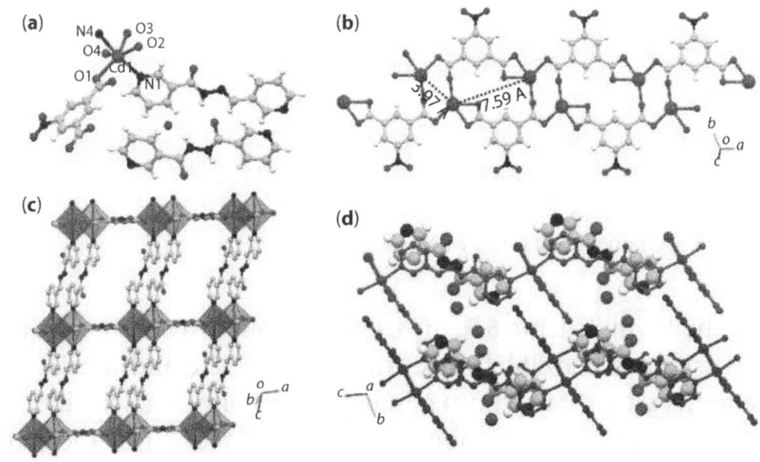

Figure 3.12 ADES-2: (a) coordination environment around Cd(II); (b) and (c) $[Cd_2(COO)_2]$ SBU in 1D metal-carboxylate chain and double lined 2D network, respectively; (d) Offset stacked 2D layers with the lattice H_2O and $\pi\cdots\pi$ stacked L [53].

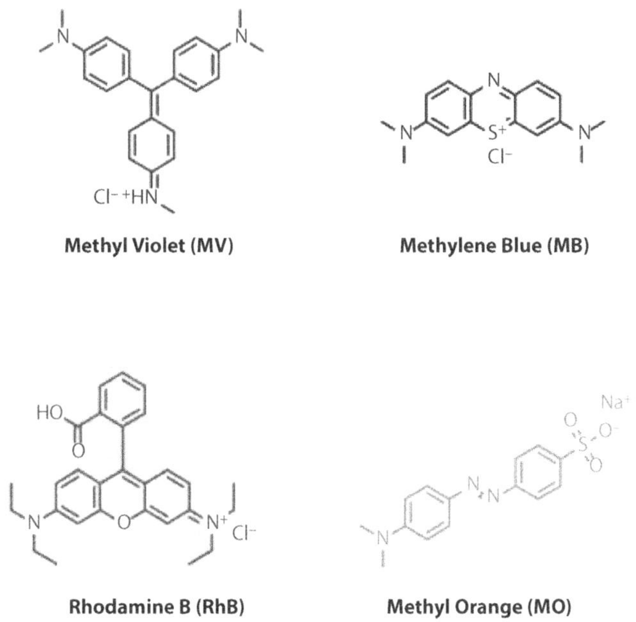

Scheme 3.2 Structures of the used organic dyes.

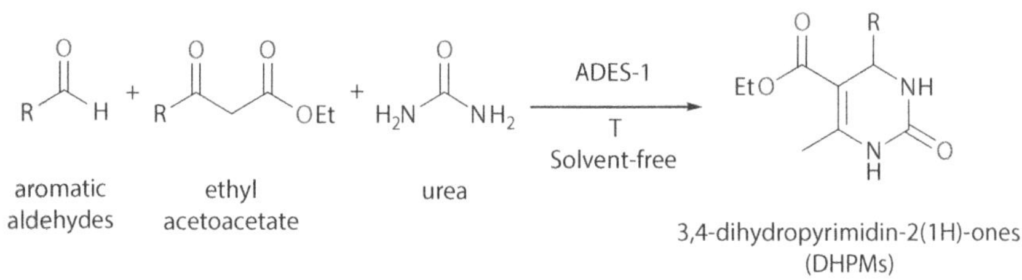

Scheme 3.3 Biginelli reaction by using ADES-1.

As shown in Scheme 3.3, the used substrates were aromatic aldehydes (such as benzaldehyde), ethyl acetoacetate and urea. In this reaction, various catalytic parameters were investigated like temperature, time, catalyst amount and solvent that all of them have been reported in Table 3.2. This reaction is performed to produce dihydropyrimidinone derivatives that are very important in pharmaceutical industry. The suggested mechanism in this catalytic reactions can be based on the production of an ionic intermediate

Table 3.2 Determination of catalytic conditions for Biginelli reaction.

Entry	Catalyst (mmol)	Temp. (°C)	Solvent	Conversion (%)	Time (min)
1	0.005	80	None	22.8	10
2	0.010	80	None	73.2	10
3	0.015	80	None	93.6	10
4	0.020	80	None	98.3	10
5	0.025	80	None	93.2	10
6	0.020	60	None	86.5	20
7	0.020	70	None	91.3	10
8	0.020	90	None	98.1	10
9	0.020	80	CH_2Cl_2	82.2	10
10	0.020	80	Toluene	76.2	120
11	0.020	80	EtOH	86.7	60
12(ADES-2)	0.020	80	None	97.4	10

(N-acyliminium) [54]. Based on the obtained findings, it is concluded that these heterogeneous MOFs structures with two different ligands (N-donor and rigid) maintained 2D network topology. The electrostatic and supramolecular interactions, Lewis acids (metal site), Lewis basic centers and polar group of nitro are responsible of their abilities.

So, the presence of functional options in the structure of these multi-functional MOFs led to the improvement of their performance. Morris *et al.* also reported one of these kinds of mixed-linker MOFs that was used as catalyst in the photoelectrochemical alcohol oxidation (Figure 3.13). This system was consisted of a Ru(L1L2)-based photosentisizer, Ru(L1L3)-based catalyst that L1 to L3 were L1: 5,5′-dicarboxy-2,2′-bipyridine, L2: 2,2'-bipyridine and L3: 2,2′:6′,2″-terpyridine along with TiO_2 and FTO. The main body of this MOF-based solid was UiO-67 framework that two Ru-based

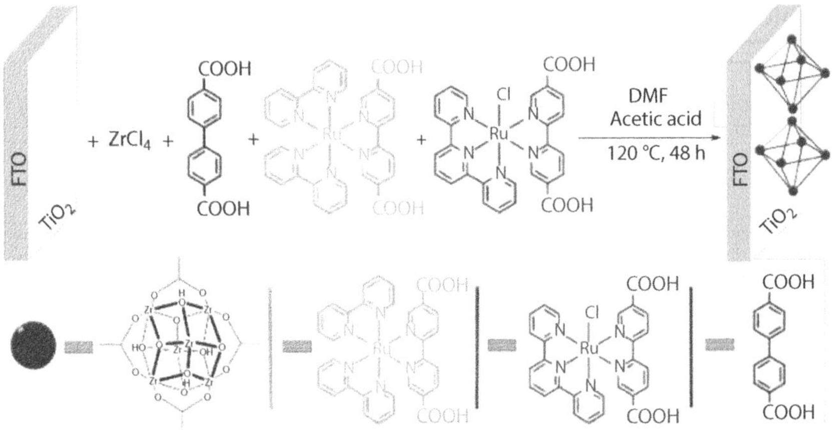

Figure 3.13 Direct synthesis of mixed-ligands MOF film, RuB-RuTB-UiO-67/TiO_2/FTO through solvothermal method [55].

compounds mentioned above were incorporated on it (RuB and RuTB are photosentisizer and catalyst, respectively).

Briefly, it should be said that this system could demonstrate suitable Faradaic efficiency (also called faradaic efficiency, faradaic yield, coulombic efficiency or current efficiency). This parameter illustrates efficiency with which charge (electrons) is transferred in a system facilitating an electrochemical process. Given, $Ru^{III}/^{II}B$ redox showed better performance in comparison to $Ru^{III}/^{II}TB$, so electron transfer happened from catalyst to photosensitizer, simply. Of course, thermodynamically-favorable mechanism, design of photosensitizer-catalyst system and the existence of cooperative effects in system should not be forgotton that they improved photocatalysis [55]. As results, mixed-components can show the considerable similar to single-component.

3.2 Conclusion

Modification of MOFs framework with mixed ligands and their variability show the impressive difference than primitive structure that can lead to complexity in the second structure. Degree of complexity in MOF is related to the structural components. One of the important ways to construct new complex MOFs is using mixed ligands. There are different kinds of mixed-linker MOFs. Due to the variability in used ligands, we observe the constructed MOFs with intriguing features that in the following the other chapters have been shown. Indeed, the complexity in the multi-component porous solids is a challenging topic that can be generated through the structural design during the synthesis methods.

References

1. Kong, X., Deng, H., Yan, F., Kim, J., Swisher, J.A., Smit, B., Yaghi, O.M., Reimer, J.A., Mapping of functional groups in metal-organic frameworks. *Science*, *341*, 6148, 882–885, 2013.
2. Bunck, D.N. and Dichtel, W.R., Mixed linker strategies for organic framework functionalization. *Chem.-Eur. J.*, *19*, 3, 818–827, 2013.
3. Zarekarizi, F. and Morsali, A., Dimension Control in Mixed Linker Metal–Organic Frameworks *via* Adjusting the Linker Shapes. *Inorg. Chem.*, *59*, 5, 2988–2996, 2020.
4. Ortín-Rubio, B., Ghasempour, H., Guillerm, V., Morsali, A., Juanhuix, J., Imaz, I., Maspoch, D., Net-Clipping: An Approach to Deduce the Topology of Metal-Organic Frameworks Built with Zigzag Ligands. *J. Am. Chem. Soc.*, 142, *20*, 9135–9140, 2020.
5. Yin, Z., Zhou, Y.-L., Zeng, M.-H., Kurmoo, M., The concept of mixed organic ligands in metal–organic frameworks: design, tuning and functions. *Dalton Trans.*, *44*, 12, 5258–5275, 2015. Top of Form.
6. Deng, H., Doonan, C.J., Furukawa, H., Ferreira, R.B., Towne, J., Knobler, C.B., Wang, B., Yaghi, O.M., Multiple functional groups of varying ratios in metal-organic frameworks. *Science*, *327*, 5967, 846–850, 2010.
7. Pang, J., Yuan, S., Qin, J., Wu, M., Lollar, C.T., Li, J., Huang, N., Li, B., Zhang, P., Zhou, H.-C., Enhancing Pore-Environment Complexity Using a Trapezoidal Linker: Toward Stepwise Assembly of Multivariate Quinary Metal–Organic Frameworks. *J. Am. Chem. Soc.*, *140*, 39, 12328–12332, 2018.

8. Seki, K., Design of an adsorbent with an ideal pore structure for methane adsorption using metal complexes. *Chem. Commun.*, 16, 1496–1497, 2001.
9. Ryu, U.J., Kim, S.J., Lim, H.-K., Kim, H., Choi, K.M., Kang, J.K., Synergistic interaction of Re complex and amine functionalized multiple ligands in metal-organic frameworks for conversion of carbon dioxide. *Sci. Rep.*, *7*, 1, 612, 2017.
10. Shultz, A.M., Farha, O.K., Adhikari, D., Sarjeant, A.A., Hupp, J.T., Nguyen, S.T., Selective surface and near-surface modification of a noncatenated, catalytically active metal-organic framework material based on Mn(salen) struts. *Inorg. Chem.*, *50*, 8, 3174–3176, 2011.
11. Sun, K., Li, L., Yu, X., Liu, L., Meng, Q., Wang, F., Zhang, R., Functionalization of mixed ligand metal-organic frameworks as the transport vehicles for drugs. *J. Colloid Interface Sci.*, *486*, 128–135, 2017.
12. Taddei, M., Costantino, F., Ienco, A., Comotti, A., Dau, P.V., Cohen, S.M., Synthesis, breathing, and gas sorption study of the first isoreticular mixed-linker phosphonate based metal–organic frameworks. *Chem. Commun.*, *49*, 13, 1315–1317, 2013.
13. Liu, B., Sun, C., Chen, G., Molecular simulation studies of separation of CH_4/H_2 mixture in metal-organic frameworks with interpenetration and mixed-ligand. *Chem. Eng. Sci.*, *66*, 13, 3012–3019, 2011.
14. Celebi, N., Aydin, M.Y., Soysal, F., Yıldız, N., Salimi, K., Core/Shell PDA@ UiO-66 Metal–Organic Framework Nanoparticles for Efficient Visible-Light Photodegradation of Organic Dyes. *ACS Appl. Nano Mater.*, *3*, 11, 11543–11554, 2020.
15. Paluka, V., Maihom, T., Probst, M., Limtrakul, J., Dehydrogenation of Ethanol to Acetaldehyde with Nitrous Oxide over the Metal-Organic Framework NU-1000: A Density Functional Theory Study. *Phys. Chem. Chem. Phys.*, *22*, 13622–13628, 2020.
16. Furukawa, H., Gándara, F., Zhang, Y.-B., Jiang, J., Queen, W.L., Hudson, M.R., Yaghi, O.M., Water adsorption in porous metal–organic frameworks and related materials. *J. Am. Chem. Soc.*, *136*, 11, 4369–4381, 2014.
17. Cliffe, M.J., Wan, W., Zou, X., Chater, P.A., Kleppe, A.K., Tucker, M.G., Wilhelm, H., Funnell, N.P., Coudert, F.-X., Goodwin, A.L., Correlated defect nanoregions in a metal–organic framework. *Nat. Commun.*, 5, 4176, 2014.
18. Pratik, S.M., Gagliardi, L., Cramer, C.J., Engineering Electrical Conductivity in Stable Zirconium-Based PCN-222 MOFs with Permanent Mesoporosity. *Chem. Mater.*, *32*, 14, 6137–6149, 2020.

19. Hong, S.W., Paik, J.W., Seo, D., Oh, J.M., Jeong, Y.K., Park, J.K., Substrate templated synthesis of single-phase and uniform Zr-porphyrin-based metal–organic frameworks. *Inorg. Chem. Front.*, *7*, 1, 221–231, 2020.
20. Jiang, H.-L., Feng, D., Wang, K., Gu, Z.-Y., Wei, Z., Chen, Y.-P., Zhou, H.-C., An exceptionally stable, porphyrinic Zr metal–organic framework exhibiting pH-dependent fluorescence. *J. Am. Chem. Soc.*, *135*, 37, 13934–13938, 2013.
21. Yuan, S., Zou, L., Qin, J.-S., Li, J., Huang, L., Feng, L., Wang, X., Bosch, M., Alsalme, A., Cagin, T., Construction of hierarchically porous metal–organic frameworks through linker labilization. *Nat. Commun.*, *8*, 15356, 2017.
22. Marx, S., Kleist, W., Baiker, A., Synthesis, structural properties, and catalytic behavior of Cu-BTC and mixed-linker Cu-BTC-PyDC in the oxidation of benzene derivatives. *J. Catal.*, *281*, 1, 76–87, 2011.
23. Kozachuk, O., Luz, I., Llabres i Xamena, F.X., Noei, H., Kauer, M., Albada, H.B., Bloch, E.D., Marler, B., Wang, Y., Muhler, M., Multifunctional, defect-engineered metal–organic frameworks with ruthenium centers: sorption and catalytic properties. *Angew. Chem. Int. Ed.*, *53*, 27, 7058–7062, 2014.
24. Yuan, S., Chen, Y.-P., Qin, J.-S., Lu, W., Zou, L., Zhang, Q., Wang, X., Sun, X., Zhou, H.-C., Linker installation: engineering pore environment with precisely placed functionalities in zirconium MOFs. *J. Am. Chem. Soc.*, *138*, 28, 8912–8919, 2016.
25. Zhang, Y.B., Zhang, W.X., Feng, F.Y., Zhang, J.P., Chen, X.M., A highly connected porous coordination polymer with unusual channel structure and sorption properties. *Angew. Chem. Int. Ed.*, *48*, 29, 5287–5290, 2009.
26. Zhang, X., Frey, B.L., Chen, Y.-S., Zhang, J., Topology-Guided Stepwise Insertion of Three Secondary Linkers in Zirconium Metal–Organic Frameworks. *J. Am. Chem. Soc.*, *140*, 24, 7710–7715, 2018.
27. Bhattacharya, B., Maity, D.K., Layek, A., Jahiruddin, S., Halder, A., Dey, A., Ghosh, S., Chowdhury, C., Datta, A., Ray, P.P., Multifunctional mixed ligand metal organic frameworks: X-ray structure, adsorption, luminescence and electrical conductivity with theoretical correlation. *CrystEngComm*, *18*, 30, 5754–5763, 2016.
28. Wang, C.-C., Ke, S.-Y., Cheng, C.-W., Wang, Y.-W., Chiu, H.-S., Ko, Y.-C., Sun, N.-K., Ho, M.-L., Chang, C.-K., Chuang, Y.-C., Four mixed-ligand Zn (II) three-dimensional metal-organic frameworks: Synthesis, structural diversity, and photoluminescent property. *Polymers*, *9*, 12, 644, 2017.

29. Sun, K., Li, L., Yu, X., Liu, L., Meng, Q., Wang, F., Zhang, R., Functionalization of mixed ligand metal-organic frameworks as the transport vehicles for drugs. *J. Colloid Interface Sci.*, *486*, 128–135, 2017.
30. Sun, Y., Ma, R., Wang, F., Guo, X., Sun, S., Guo, H., Alexandrov, E.V., Design, Two Novel Self-Catenated Metal–Organic Frameworks with Large Accessible Channels Obtained by a Mixed-Ligand Strategy: Adsorption of Dichromate and Ln^{3+} Postsynthetic Modification. *Cryst. Growth Des.*, *19*, 9, 5267–5274, 2019.
31. Yu, L.-G., Sun, Y.-G., Wang, Z.-L., Mixed-ligand strategy affording two new metal-organic frameworks: Photocatalytic, luminescent and anti-lung cancer properties. *J. Mol. Struct.*, *1180*, 209–214, 2019.
32. Qiu, Y.-C., Yuan, S., Li, X.-X., Du, D.-Y., Wang, C., Qin, J.-S., Drake, H.F., Lan, Y.-Q., Jiang, L., Zhou, H.-C., Face-Sharing Archimedean Solids Stacking for the Construction of Mixed-Ligand Metal–Organic Frameworks. *J. Am. Chem. Soc.*, *141*, 35, 13841–13848, 2019.
33. Bae, Y.-S., Mulfort, K.L., Frost, H., Ryan, P., Punnathanam, S., Broadbelt, L.J., Hupp, J.T., Snurr, R.Q., Separation of CO_2 from CH4 using mixed-ligand metal–organic frameworks. *Langmuir*, *24*, 16, 8592–8598, 2008.
34. Noro, S.-i., Kitagawa, S., Wada, T., Hydrogen-bonding assemblies constructed from metalloligand building blocks and H_2O. *Inorg. Chim. Acta*, *358*, 2, 423–428, 2005.
35. An, J., Geib, S.J., Rosi, N.L., Cation-triggered drug release from a porous zinc–adeninate metal–organic framework. *J. Am. Chem. Soc.*, *131*, 24, 8376–8377, 2009.
36. Sudik, A.C., Côté, A.P., Wong-Foy, A.G., O'Keeffe, M., Yaghi, O.M.A., metal–organic framework with a hierarchical system of pores and tetrahedral building blocks. *Angew. Chem. Int. Ed.*, *45*, 16, 2528–2533, 2006.
37. Lin, T.-R., Lee, C.-H., Lan, Y.-C., Mendiratta, S., Lai, L.-L., Wu, J.-Y., Chi, K.-M., Lu, K.-L., Paddlewheel SBU based Zn MOFs: Syntheses, Structural Diversity, and CO2 Adsorption Properties. *Polymers*, *10*, 12, 1398, 2018.
38. Hu, X.-L., Liu, F.-H., Wang, H.-N., Qin, C., Sun, C.-Y., Su, Z.-M., Liu, F.-C., Controllable synthesis of isoreticular pillared-layer MOFs: gas adsorption, iodine sorption and sensing small molecules. *J. Mater. Chem. A*, *2*, 36, 14827–14834, 2014.
39. Kleist, W., Jutz, F., Maciejewski, M., Baiker, A., Mixed-Linker Metal-Organic Frameworks as Catalysts for the Synthesis of Propylene

Carbonate from Propylene Oxide and CO2. *Eur. J. Inorg. Chem., 2009*, 24, 3552–3561, 2009.

40. Zhang, L., Wu, H.B., Lou, X.W., Metal–organic-frameworks-derived general formation of hollow structures with high complexity. *J. Am. Chem. Soc., 135*, 29, 10664–10672, 2013.
41. Liu, L., Konstas, K., Hill, M.R., Telfer, S.G., Programmed pore architectures in modular quaternary metal–organic frameworks. *J. Am. Chem. Soc., 135*, 47, 17731–17734, 2013.
42. Dutta, A., Wong-Foy, A.G., Matzger, A.J., Coordination copolymerization of three carboxylate linkers into a pillared layer framework. *Chem. Sci., 5*, 10, 3729–3734, 2014.
43. Thompson, J.A., Blad, C.R., Brunelli, N.A., Lydon, M.E., Lively, R.P., Jones, C.W., Nair, S., Hybrid zeolitic imidazolate frameworks: controlling framework porosity and functionality by mixed-linker synthesis. *Chem. Mater., 24*, 10, 1930–1936, 2012.
44. Xuan, W., Ye, C., Zhang, M., Chen, Z., Cui, Y., A chiral porous metallosalan-organic framework containing titanium-oxo clusters for enantioselective catalytic sulfoxidation. *Chem. Sci., 4*, 8, 3154–3159, 2013.
45. Habib, H.A., Sanchiz, J., Janiak, C., Mixed-ligand coordination polymers from 1, 2-bis (1, 2, 4-triazol-4-yl) ethane and benzene-1, 3, 5-tricarboxylate: Trinuclear nickel or zinc secondary building units for three-dimensional networks with crystal-to-crystal transformation upon dehydration. *Dalton Trans.*, 13, 1734–1744, 2008.
46. Bi, X., Zhang, Y.A., Zhang, F., Zhang, S., Wang, Z., Jin, J., MOF Nanosheet-Based Mixed Matrix Membranes with Metal–Organic Coordination Interfacial Interaction for Gas Separation. *ACS Appl. Mater. Interfaces, 12*, 43, 49101–49110, 2020.
47. Bruno, R., Mon, M., Escamilla, P., Ferrando-Soria, J., Esposito, E., Fuoco, A., Monteleone, M., Jansen, J.C., Elliani, R., Tagarelli, A., Armentano, D., Bioinspired Metal-Organic Frameworks in Mixed Matrix Membranes for Efficient Static/Dynamic Removal of Mercury from Water. *Adv. Funct. Mater., 31*, 6, 2008499, 2021.
48. Henninger, S.K., Habib, H.A., Janiak, C., MOFs as adsorbents for low temperature heating and cooling applications. *J. Am. Chem. Soc., 131*, 8, 2776–2777, 2009.
49. Jeremias, F., Khutia, A., Henninger, S.K., Janiak, C., MIL-100 (Al, Fe) as water adsorbents for heat transformation purposes—a promising application. *J. Mater. Chem., 22*, 20, 10148–10151, 2012.
50. Seo, Y.K., Yoon, J.W., Lee, J.S., Hwang, Y.K., Jun, C.H., Chang, J.S., Wuttke, S., Bazin, P., Vimont, A., Daturi, M., Energy-efficient

dehumidification over hierachically porous metal–organic frameworks as advanced water adsorbents. *Adv. Mater.*, *24*, 6, 806–810, 2012.
51. Li, J.R., Timmons, D.J., Zhou, H.C., Interconversion between molecular polyhedra and metal– organic frameworks. *J. Am. Chem. Soc.*, *131*, 18, 6368–6369, 2009.
52. Son, H.J., Jin, S., Patwardhan, S., Wezenberg, S.J., Jeong, N.C., So, M., Wilmer, C.E., Sarjeant, A.A., Schatz, G.C., Snurr, R.Q., Farha, O.K., Light-harvesting and ultrafast energy migration in porphyrin-based metal–organic frameworks. *J. Am. Chem. Soc.*, *135*, 2, 862–869, 2013.
53. Patel, U., Parmar, B., Patel, P., Dadhania, A., Suresh, E., The synthesis and characterization of Zn (ii)/Cd (ii) based MOFs by a mixed ligand strategy: a Zn (ii) MOF as a dual functional material for reversible dye adsorption and as a heterogeneous catalyst for the Biginelli reaction. *Mater. Chem. Front.*, *5*, 1, 304–314, 2021.
54. Kappe, C.O., A reexamination of the mechanism of the Biginelli dihydropyrimidine synthesis. Support for an N-Acyliminium ion intermediate. *J. Org. Chem.*, *62*, 21, 7201–7204, 1997.
55. Lin, S., Cairnie, D.R., Davis, D., Chakraborty, A., Cai, M., Morris, A.J., Photoelectrochemical alcohol oxidation by mixed-linker metal–organic frameworks. *Faraday Discuss.*, *225*, 371–383, 2021.

4

Complexity Based on Ligand—Part 2

Abstract
In most cases, organic ligands that are used in the creation of MOFs are ditopic (O-donor ligands such as carboxylates or N-donor ligands like pyridine pillars). But, recently, organic linkers are employed in large varieties like polytopic (tri- and tetratopic) and multi-heterotopic (toward multiple coordination of metal ions by using mixed N,O donors) ligands. They are factors that complexity of metal-organic frameworks arises from both of them. In this chapter, we present some of these ligands along with several examples for various applications.

Keywords: Multi-heterotopic ligand, polytopic organic linker, mixed donor ligands, symmetry, tri- and tetratopic linkers

4.1 Polytopic Linkers

One of the complicated cases of MOFs is the using kinds of linkers in designing especially polytopic forms. Kind of employed linkers along with the other components can affect MOFs complexity. Generally, a large number of crystalline MOFs were synthesized with metal sites and simple bridging ligands especially ditopic carboxylate linkers. The linkers are not limited to carboxylate but also the other linkers are utilized as phosphonate, pyridine, and azolate [1–3], too. Many researchers focused chiefly on all of the chemical properties of MOFs from the components to the description of their designing, topologies, and kind of crystal nets. There are many MOFs that can be formed by various polytopic organic linkers as tritopic, tetratopic, hexatopic, and octatopic. Metal units can be joined by using these organic linkers, and then, periodic frameworks can

Ali Morsali and Kayhaneh Berijani. Metal-Organic Frameworks with Heterogeneous Structures, (57–70) © 2021 Scrivener Publishing LLC

be produced. Also, infinite SBUs in MOF crystalline structure can embed less-symmetric polytopic linkers [4]. Design of this kind of MOFs has been considerably attracted from the point of potentially features. The type of nets in MOFs frameworks with polytopic linkers can be diverse and they can induce complexity to MOFs final structure. UTSA-30 ($[Yb(-COO)_3]_\infty$) as a rod MOF has been known with tritopic linker. The generated net is hyb that in this rod net, and the considered complexity appears [5]. Many MOFs were built with various tritopic linkers and different metal node as paddle wheel, octahedral, trigonal prismatic, and multiple SBUs. The distorted geometry, low symmetry, and different length kinds of function of components will generate complex environment in the framework. A series of tritopic linkers has been shown in Figure 4.1.

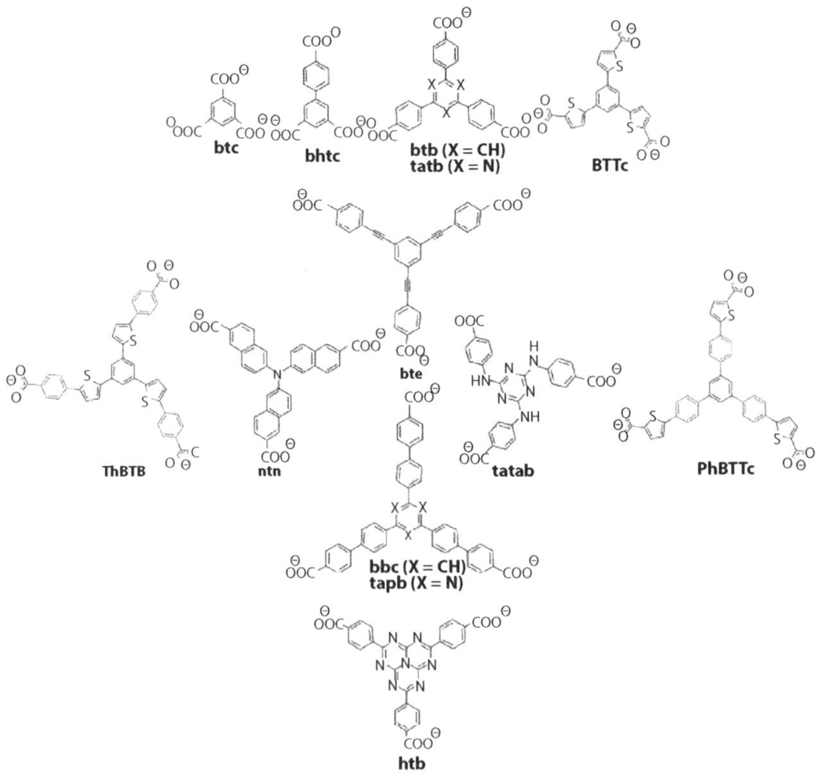

Figure 4.1 Some of the tritopic linkers.

Employing various linkers with considerable differences leads to nets with distinct topologies and SBUs that can twist from flat case. The steric hindrance in net structure can also produce twist angles so high complexity can be created. With the further studies, it is expected that the symmetry of MOF structure should be reduced to increase complexity by using the asymmetric or complicated organic linkers. By replacing a primitive linker with a new polytopic linker, the symmetry can be reduced though crystalline structure is maintained (Figure 4.2) [6]. In these MOFs, complex cavity can be produced with different size that can cause the increasing of complexity.

Interestingly, MOFs were also reported with mixed ligands along with polytopic linkers and significant improvement in the structure and design. They have better intrinsic properties than their counterparts with single ligand. Utilization of tritopic and ditopic carboxylates, carboxylate and pyridine, and polytopic N-heterocyclic linkers leads to plentiful mixed ligand-based MOFs.

Some examples of tetratopic linkers have been shown in Figure 4.3. The coordination polymers can be interesting much as sensors due to the use of various N-donor ligands and polycarboxylate with 1D, 2D, and 3D structures. The diversity in these kinds of frameworks happened through steric hindrance of the different positions of the employed linkers. In addition, their luminescent properties showed that these compounds had high selectivity and sensitivity than detection of some ions as Fe^{3+}, $Cr_2O_7^{2-}$/CrO_4^{2} in aqueous solution [7].

In the past, Yaghi and his groups investigated the direct synthesis of a multivariate MOF (MTV-MOF) by eight different derivatives of one type of linker that were incorporated into MOF-5 framework.

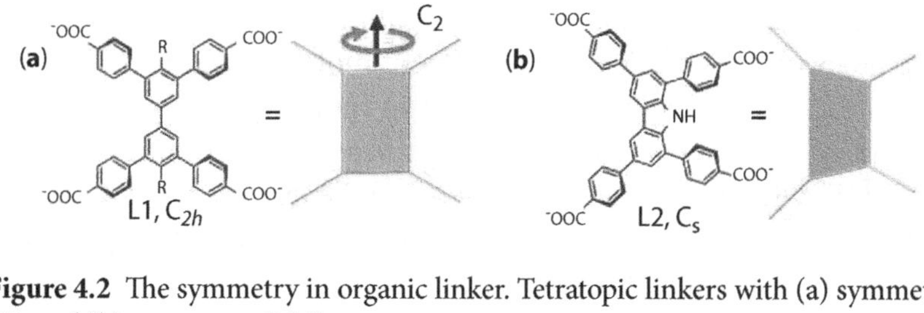

Figure 4.2 The symmetry in organic linker. Tetratopic linkers with (a) symmetry $C2_h$ and (b) symmetry C_s [6].

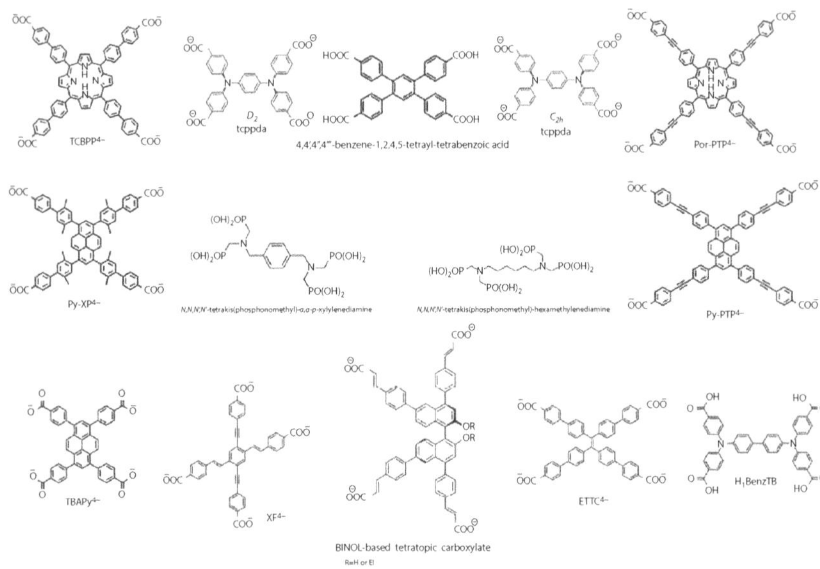

Figure 4.3 Some of the tetratopic linkers.

It was valuable that the formed pores with functional groups have complexity due to the existence of different functions. The presence of functional groups in the used linkers structure is an important method to generate complexity with various properties in MOFs with mixed ligand or polytopic linkers [8]. However, post-synthetic modification (PSM) is a strategy for the functionalization of MOFs because direct synthesis may be more difficult and generates complexity through steric interactions due to bulky substituents. As explained, mixed ligands in either mixed ditopic linkers or mixed ligands case, to form the mixture of polytopic-ditopic linkers, can affect the diverse features of MOFs. For example, the capability of the adsorption of gases with remarkable values in MOFs due to specific interactions is valuable. With this ability and difference of standard adsorption enthalpy of the collection of gases, some MOFs can separate gases. It is worth noting that in mixed-ligand structure the content of suitable gas adsorption, whereas this amount in the parent MOF with single ligand is lower. From these results, it could be concluded which the modification of porous coordination polymers can affect their adsorption properties [9].

A new mixed-ligand MOF was synthesized as the ordered light-harvesting system (MLMs) that was prepared from the combination of Zr_6 cluster and suitable ratio of H_4TBAPy and Zn-TCPP during solvothermal reactions (Figure 4.4). Applied ligands with their symmetry and connectivity cause the great overlap to happen between the emission and absorption of pyrene and porphyrin, respectively. This event is ideal and very good for the efficient energy transfer (EnT) process in MLMs because EnT is a necessary action in photosynthesis. Moreover, the synthesized pyrene and porphyrin-based MLMs can be used as superior photo-induced singlet oxygen generators that this ability is due to the absorption and energy transfer between pyrene and porphyrin ligands which are done completely [10]. In addition to the flat tetratopic linkers, tetrahedral organic linkers can be used in the construction of MOF structures with diversity, too [11–14]. By contrast flat tetratopic linkers, these kinds of linkers much less are known because their synthesis is challenging [15]. With development of MOFs world, the efficacy of correct choosing in the design is clear. The design of new organic linkers has important role in the creation of reticular materials as MOFs with intricate structural complexity. A series of hexatopic and

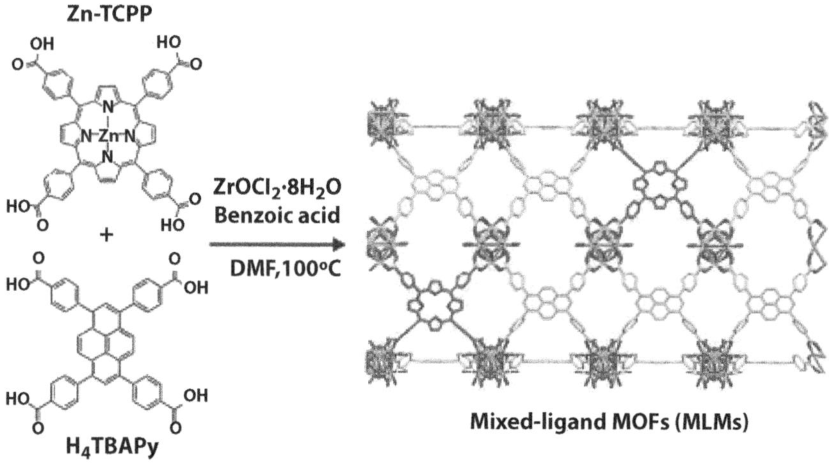

Figure 4.4 A mixed-ligand MOF with tetratopic organic linkers [10].

octatopic linkers were also reported with carboxylate head that some of them have been illustrated in Figure 4.5.

The metal units can be linked to hexatopic linkers and expansion of crystalline net that leads to the construction of 3D MOFs with high surface area and large pores.

Undoubtedly, this kind of linkers as 5,5',5"-[1,3,5-benzenetriyltris (carbonylimino)] tris-1,3-benzenedicarboxylate can produce supermolecular building block. These linkers can be semi-rigid with low symmetry so they can induce complexity into MOFs framework. Several MOFs were synthesized by dinuclear paddlewheel units and

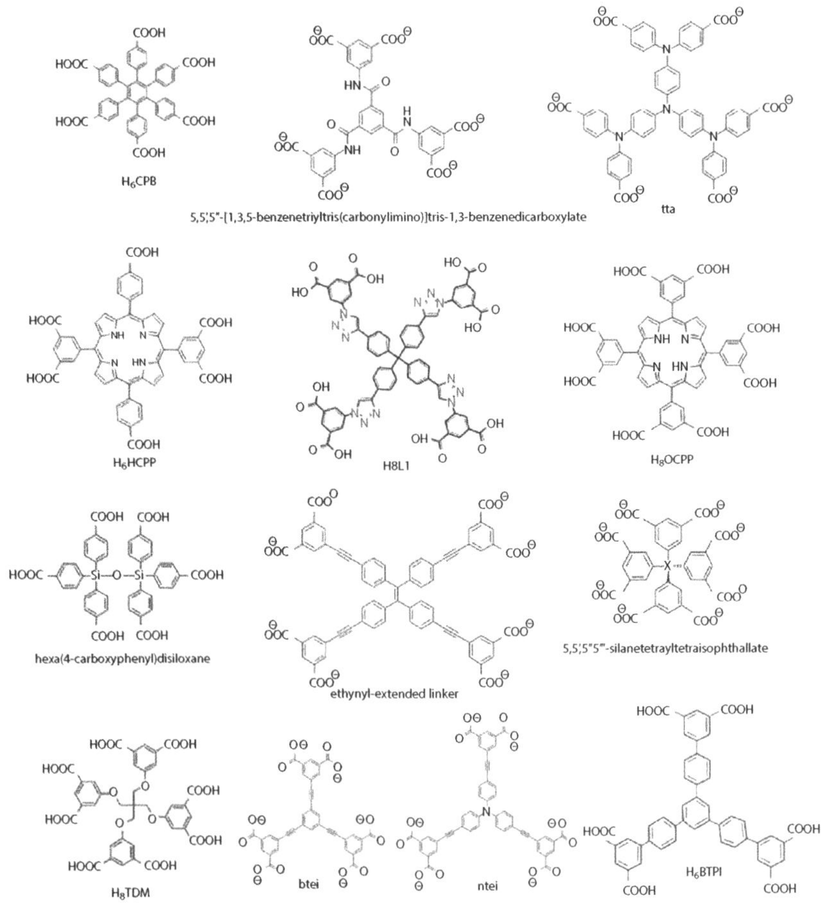

Figure 4.5 Some of the multi-topic linkers.

hexatopic linkers. The presence of microwindows in these MOFs can link mesocavities together. With this strategy, the interpenetration of MOFs cannot happen *via* the produced network and the adsorption amount is several times higher. Shape, size, and chemical properties of the used components particularly organic linkers have supplied possibilities in MOFs applications such as selective catalysis and adsorption-separation and especially in gas or drug storage [16–24]. Geometry of MOFs can lead to the construction of MOFs with high stability, large cavities, and high performance in the various applications [25]. The high tunability in MOFs provides conditions that large organic linkers can be used as octatopic linkers.

4.2 Multi-Heterotopic Ligands

As known, metal-organic frameworks are crystalline frameworks, composed of inorganic clusters and organic linkers. Most of the used organic linkers in MOF synthesis are highly symmetric, leading to architectures with only a few network topologies, aiming network transitivity with high performance. The tunability of MOFs structure and their chemical/physical properties lead to synthesis of MOFs that are excellent candidates in different scientific fields. There are several valid methods for further processing in design of MOFs materials. Up to now, numerous researchers have performed investigations on porous MOFs with structural diversity. But, designings are not limited to usual metal nodes and homotopic organic linkers with high symmetry [26, 27]. In contrast to this simplicity, step by step, the considerable progressing is observed in MOFs with rich chemistry and outstanding properties [28]. The development of the design of linkers with heteroatoms and using them in the construction of MOFs has attracted much attention. In this section, we speak about MOFs with multi-heterotopic linkers; however, there are not many of these kinds of MOFs due to their high complexity. There are reports about porous MOFs that have one kind of linker with two different coordinating groups. MOFs were reported with mixed donors in organic linkers as N,O,S that can provide active sites for different applications such as 5-(1H-imidazol-1-yl)isophthalic acid [29], 6-mercaptonicotinic acid [30], 5-hydroxybenzene-1,3-dicarboxylic acid [31], and 4-((3,5-dimethyl-1H-pyrazol-4-yl)methyl)-benzoic acid [32]. But, it seems that the constructing multi-heterotopic ligands can be difficult so they have not been reported much. For the first time, Yaghi and coworkers reported the synthesis of a Zn-based MOF (MOF-910) by using a unique multi-heterotopic ligand with three different kinds of dentates. This stable and porous MOF with BET surface area 2140 m^2g^{-1} crystallizes in trigonal space group *R-3c* with specific lattice parameters (a: 47.239(2) Å; c: 27.1216(12) Å). Its structure has new topology (tto: triangles, tetrahedra, octahedra) with hexagonal channels that has been shown in Figure 4.6. The phenylyne-1-benzoate, 3-benzosemiquinonate, and 5-oxidopyridine (PBSP) are three distinct heads of considered heterotopic linkers. They have not high symmetry and their point group is C_1.

The form of designed SBUs is rods that finally lead to crystalline 3D helical rod MOF which may not have interpenetration. Most of rod SBUs induce complicated nature to MOFs net [33, 34]. Of course, there are some examples of rod MOFs that have remarkable complexity in their nets [35]. In this 3D framework, one dimension SBUs generate the hexagonal channels and disorders are seen in the benzoate moieties. Using the multi-heterotritopic organic linkers with less symmetry to coordinate metallic sites in MOFs structure can be an innovation in the creation of noncentrosymmetric MOFs. Surely, the control of selectivity and helicity to SBUs is very significant which influences on the degree of complexity. Probably, in this kind of MOFs with the existence of helicity, chiral property is seen.

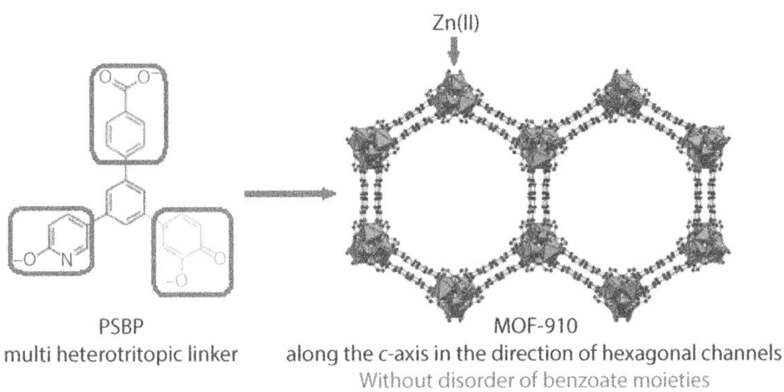

Figure 4.6 A kind of MOF with multi-hetero topic organic linker [4].

So, they can have asymmetric applications with the breathing functionality [36–39].

Given that, MOFs with heterogeneous structure can be effective like MOFs with homogeneous structure, we want to present another example. In a new report published by Zheng *et al.*, SCNU-Z4 was presented as a MOF with suitable cage cavity and channels [40]. Two main constituents of body are Cu(II) bifunctional tripodal nitrogen-donor ligand. It is interesting that this framework with high chemical stability could be used to remove I_2 molecules and hazardous dyes from solution. This process is very important because these compounds are considered as environmental contaminants. There are two kinds of SBU in this structure, trinuclear and tetranuclear SBUs (Figure 4.7). Interestingly, multiporosity in MOF has been observed owing to the presence of two kinds of pores

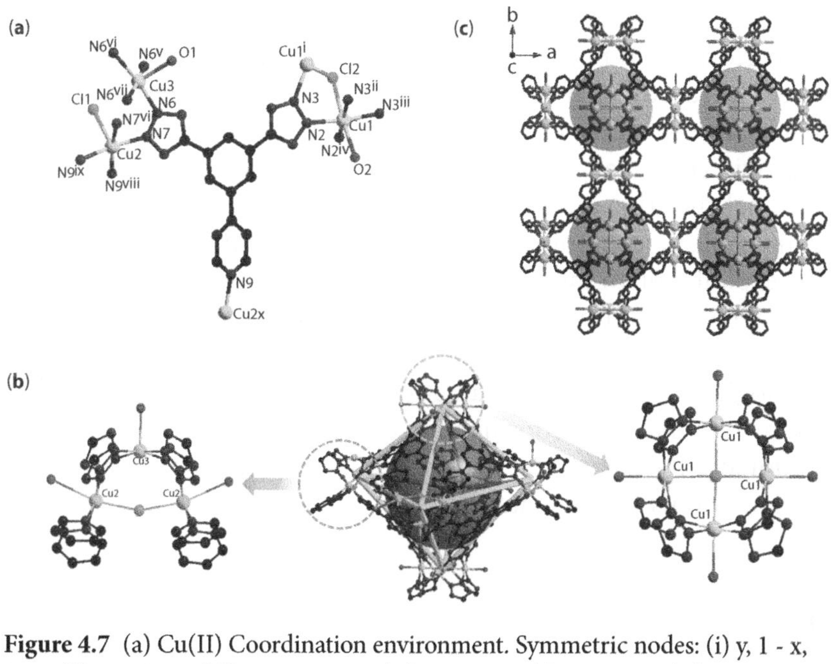

Figure 4.7 (a) Cu(II) Coordination environment. Symmetric nodes: (i) y, 1 - x, 1 - z; (ii) y, x, 1 – z; (iii) 1 - y, x, 1 – z; (iv) 1 – x, y, z; (v) x, 1 – y, z; (vi) -x, 1 – y, z; (vii) -x, y, z; (viii) y, x, -z; (ix) -y, x, -z; (x) y, -x, -z. (b) Distorted truncated octahedron cage: (trinuclear + tetranuclear) SBUs. (c) Illustration of three dimensional porous framework in the considered MOF [40].

in MOF, micropores and mesopores that have been confirmed by BET. This property indicated ability to separate the CO_2/N_2 gases. Because, as known, porous materials can significantly improve the selectivity of CO_2 over N_2. Uncoordinated N donor, hydrolytic stability, hydrogen bonding and π…π interactions, structural defect, diverse pores and surface area are effective agents in the improvement of framework performance.

4.3 Conclusion

The multidentate linkers can be used to prepare robust MOFs with the different topological structures and they can have several porosities, simultaneously. These designs are promising methods to create unique MOFs with interesting framework and features. The accessible open channels in these MOFs by using polytopic ligands can be employed in the different applications like catalytic reactions which are performed in pores. To synthesis this kind of ligands, there are several methods such as Sonogashira-type cross-coupling reactions that have been used in the synthesis of polytopic ligands that consist of a rigid central arylene platform, ethynylene spacer subunits, and terminal chelating picolinate subunits [41]. The syntheses take place from available compounds in only two steps and very good yields. The exploitation of these interesting organic ligands in MOFs is now in progress. About multi-hetero topic ligands, design of this kind of linkers shows that we can achieve to crystalline MOFs with intricate topologies by using the organic linkers with lower regularity (low symmetry) that their coordinating atoms are not identical. The formation of MOFs based on multi-heterotopic ligands is an important level of porous MOFs materials.

References

1. Wharmby, M.T., Mowat, J.P., Thompson, S.P., Wright, P.A., Extending the pore size of crystalline metal phosphonates toward the mesoporous regime by isoreticular synthesis. *J. Am. Chem. Soc.*, *133*, 5, 1266–1269, 2011.

2. Zarekarizi, F., Morsali, A., Buyukgungor, O., Rapid and Selective Water Remediation through a Functionalized Pillar's Core of a Novel Metal-Organic Framework. *Cryst. Growth Des., 20*, 9, 6109–6116, 2020.
3. Dinca, M., Yu, A.F., Long, J.R., Microporous metal-organic frameworks incorporating 1, 4-benzeneditetrazolate: Syntheses, structures, and hydrogen storage properties. *J. Am. Chem. Soc., 128*, 27, 8904–8913, 2006.
4. Catarineu, N.R., Schoedel, A., Urban, P., Morla, M.B., Trickett, C.A., Yaghi, O.M., Two principles of reticular chemistry uncovered in a metal-organic framework of heterotritopic linkers and infinite secondary building units. *J. Am. Chem. Soc., 138*, 34, 10826–10829, 2016.
5. He, Y., Xiang, S., Zhang, Z., Xiong, S., Fronczek, F.R., Krishna, R., O'Keeffe, M., Chen, B., A microporous lanthanide-tricarboxylate framework with the potential for purification of natural gas. *Chem. Commun., 48*, 88, 10856–10858, 2012.
6. Pang, J., Yuan, S., Qin, J., Wu, M., Lollar, C.T., Li, J., Huang, N., Li, B., Zhang, P., Zhou, H.-C., Enhancing Pore-Environment Complexity Using a Trapezoidal Linker: Toward Stepwise Assembly of Multivariate Quinary Metal-Organic Frameworks. *J. Am. Chem. Soc., 140*, 39, 12328–12332, 2018.
7. Lin, Y., Zhang, X., Chen, W., Shi, W., Cheng, P., Three cadmium coordination polymers with carboxylate and pyridine mixed ligands: luminescent sensors for FeIII and CrVI ions in an aqueous medium. *Inorg. Chem., 56*, 19, 11768–11778, 2017.
8. Deng, H., Doonan, C.J., Furukawa, H., Ferreira, R.B., Towne, J., Knobler, C.B., Wang, B., Yaghi, O.M., Multiple functional groups of varying ratios in metal-organic frameworks. *Science, 327*, 5967, 846–850, 2010.
9. Reinsch, H., Waitschat, S., Stock, N., Mixed-linker MOFs with CAU-10 structure: synthesis and gas sorption characteristics. *Dalton Trans., 42*, 14, 4840–4847, 2013.
10. Park, K.C., Seo, C., Gupta, G., Kim, J., Lee, C.Y., Efficient energy transfer (EnT) in pyrene-and porphyrin-based mixed-ligand metal-organic frameworks. *ACS Appl. Mater. Interfaces, 9*, 44, 38670–38677, 2017.
11. Zhao, D., Timmons, D.J., Yuan, D., Zhou, H.-C., Tuning the topology and functionality of metal- organic frameworks by ligand design. *Acc. Chem. Res., 44*, 2, 123–133, 2010.
12. Paz, F.A.A., Klinowski, J., Vilela, S.M., Tome, J.P., Cavaleiro, J.A., Rocha, J., Ligand design for functional metal-organic frameworks. *Chem. Soc. Rev., 41*, 3, 1088–1110, 2012.

13. O'Keeffe, M. and Yaghi, O.M., Deconstructing the crystal structures of metal–organic frameworks and related materials into their underlying nets. *Chem. Rev.*, *112*, 2, 675–702, 2011.
14. Carlucci, L., Ciani, G., Proserpio, D.M., Polycatenation, polythreading and polyknotting in coordination network chemistry. *Coord. Chem. Rev.*, *246*, 1–2, 247–289, 2003.
15. Lu, W., Wei, Z., Gu, Z.-Y., Liu, T.-F., Park, J., Park, J., Tian, J., Zhang, M., Zhang, Q., Gentle III, T., Tuning the structure and function of metal–organic frameworks via linker design. *Chem. Soc. Rev.*, *43*, 16, 5561–5593, 2014.
16. Rosi, N.L., Eckert, J., Eddaoudi, M., Vodak, D.T., Kim, J., O'Keeffe, M., Yaghi, O.M., Hydrogen storage in microporous metal-organic frameworks. *Science*, *300*, 5622, 1127–1129, 2003.
17. Zhao, D., Yuan, D., Zhou, H.-C., Science, E., The current status of hydrogen storage in metal–organic frameworks. *Energy Environ. Sci.*, *1*, 2, 222–235, 2008.
18. Eddaoudi, M., Kim, J., Rosi, N., Vodak, D., Wachter, J., O'Keeffe, M., Yaghi, O.M., Systematic design of pore size and functionality in isoreticular MOFs and their application in methane storage. *Science*, *295*, 5554, 469–472, 2002.
19. Ma, S., Sun, D., Simmons, J.M., Collier, C.D., Yuan, D., Zhou, H.-C., Metal-organic framework from an anthracene derivative containing nanoscopic cages exhibiting high methane uptake. *J. Am. Chem. Soc.*, *130*, 3, 1012–1016, 2008.
20. Ma, S., Sun, D., Wang, X.S., Zhou, H.C., A mesh-adjustable molecular sieve for general use in gas separation. *Angew. Chem. Int. Ed.*, *46*, 14, 2458–2462, 2007.
21. Horike, S., Dinca, M., Tamaki, K., Long, J.R., Size-selective Lewis acid catalysis in a microporous metal-organic framework with exposed Mn^{2+} coordination sites. *J. Am. Chem. Soc.*, *130*, 18, 5854–5855, 2008.
22. Alkordi, M.H., Liu, Y., Larsen, R.W., Eubank, J.F., Eddaoudi, M., Zeolite-like metal–organic frameworks as platforms for applications: on metalloporphyrin-based catalysts. *J. Am. Chem. Soc.*, *130*, 38, 12639–12641, 2008.
23. Horcajada, P., Serre, C., Vallet-Regí, M., Sebban, M., Taulelle, F., Férey, G., Metal–organic frameworks as efficient materials for drug delivery. *Angew. Chem. Int. Ed.*, *45*, 36, 5974–5978, 2006.
24. Liu, B., Shioyama, H., Akita, T., Xu, Q., Metal-organic framework as a template for porous carbon synthesis. *J. Am. Chem. Soc.*, *130*, 16, 5390–5391, 2008.

25. Li, P.-Z., Wang, X.-J., Liu, J., Lim, J.S., Zou, R., Zhao, Y., A triazole-containing metal–organic framework as a highly effective and substrate size-dependent catalyst for CO_2 conversion. *J. Am. Chem. Soc.*, *138*, 7, 2142–2145, 2016.
26. Abdollahi, N. and Morsali, A., Catalytic improvement by open metal sites in a new mixed-ligand hetero topic metal–organic framework. *Polyhedron*, *159*, 72–77, 2019.
27. Furukawa, H., Cordova, K.E., O'Keeffe, M., Yaghi, O.M., The chemistry and applications of metal-organic frameworks. *Science*, *341*, 6149, 1230444, 2013.
28. Jiang, H.-L., Makal, T.A., Zhou, H.-C., Interpenetration control in metal–organic frameworks for functional applications. *Coord. Chem. Rev.*, *257*, 15–16, 2232–2249, 2013.
29. Zhu, S.-L., Ou, S., Zhao, M., Shen, H., Wu, C.-D., A porous metal–organic framework containing multiple active Cu $^{2+}$ sites for highly efficient cross dehydrogenative coupling reaction. *Dalton Trans.*, *44*, 5, 2038–2041, 2015.
30. Humphrey, S.M., Chang, J.S., Jhung, S.H., Yoon, J.W., Wood, P.T., Porous Cobalt (II)–Organic Frameworks with Corrugated Walls: Structurally Robust Gas-Sorption Materials. *Angew. Chem. Int. Ed.*, *46*, 1–2, 272–275, 2007.
31. Lin, J.-D., Wu, S.-T., Li, Z.-H., Du, S.-W., A series of novel Pb (II) or Pb (II)/M (II)(M= Ca and Sr) hybrid inorganic–organic frameworks based on polycarboxylic acids with diverse Pb–O–M (M= Pb, Ca and Sr) inorganic connectivities. *CrystEngComm*, *12*, 12, 4252–4262, 2010.
32. Xiao, Q., Wu, Y., Li, M., O'Keeffe, M., Li, D., A metal–organic framework with rod secondary building unit based on the Boerdijk–Coxeter helix. *Chem. Commun.*, *52*, 77, 11543–11546, 2016.
33. Li, Y.-W., Zhao, J.-P., Wang, L.-F., Bu, X.-H., An Fe-based MOF constructed from paddle-wheel and rod-shaped SBUs involved *in situ* generated acetate. *CrystEngComm*, *13*, 20, 6002–6006, 2011.
34. Du, D.-Y., Qin, J.-S., Sun, Z., Yan, L.-K., O'keeffe, M., Su, Z.-M., Li, S.-L., Wang, X.-H., Wang, X.-L., Lan, Y.-Q., An unprecedented (3, 4, 24)-connected heteropolyoxozincate organic framework as heterogeneous crystalline Lewis acid catalyst for biodiesel production. *Sci. Rep.*, *3*, 2616, 2013.
35. Inge, A.K., Köppen, M., Su, J., Feyand, M., Xu, H., Zou, X., O'Keeffe, M., Stock, N., Unprecedented topological complexity in a metal–organic framework constructed from simple building units. *J. Am. Chem. Soc.*, *138*, 6, 1970–1976, 2016.

36. Rosi, N.L., Eddaoudi, M., Kim, J., O'Keeffe, M., Yaghi, O.M., Infinite Secondary Building Units and Forbidden Catenation in Metal-Organic Frameworks. *Angew. Chem. Int. Ed.*, *41*, 2, 284–287, 2002.
37. Horcajada, P., Salles, F., Wuttke, S., Devic, T., Heurtaux, D., Maurin, G., Vimont, A., Daturi, M., David, O., Magnier, E., How linker's modification controls swelling properties of highly flexible iron (III) dicarboxylates MIL-88. *J. Am. Chem. Soc.*, *133*, 44, 17839–17847, 2011.
38. Ma, L., Abney, C., Lin, W., Enantioselective catalysis with homochiral metal–organic frameworks. *Chem. Soc. Rev.*, *38*, 5, 1248–1256, 2009.
39. Das, M.C., Guo, Q., He, Y., Kim, J., Zhao, C.-G., Hong, K., Xiang, S., Zhang, Z., Thomas, K.M., Krishna, R., Interplay of metalloligand and organic ligand to tune micropores within isostructural mixed-metal organic frameworks (M′ MOFs) for their highly selective separation of chiral and achiral small molecules. *J. Am. Chem. Soc.*, *134*, 20, 8703–8710, 2012.
40. Wang, G.Q., Huang, J.F., Huang, X.F., Deng, S.Q., Zheng, S.R., Cai, S.L., Fan, J., Zhang, W.G., A hydrolytically stable cage-based metal–organic framework containing two types of building blocks for the adsorption of iodine and dyes. *Inorg. Chem. Front.*, *8*, 4, 1083–1092, 2021.
41. Jornet-Mollá, V. and Romero, F.M., Synthesis of rigid ethynyl-bridged polytopic picolinate ligands for MOF applications. *Tetrahedron Lett.*, *56*, 6120–6122, 2015.

5

Complexity Based on Metal Node

Abstract
Complex architectures can be observed in metal-organic frameworks (MOFs). Despite MOFs with one kind of node, mixed-metal MOFs can be also generated with additional degrees of structural heterogeneity. They include at least two different metal nodes in their bodies and they can have different applications. In this chapter, we discuss in detail complexity in MOFs based on mixed-metal and multiple SBUs (secondary building units) and also applications of them.

Keywords: Mixed-metal MOFs, multiple SBU, structural diversity, homometallic and heterometallic MOFs

5.1 Mixed Metal

As we know, in recent years, MOFs have been noticed due to the variation in their components as metal nodes or organic linkers that their nature and size can affect the structure designing, features, and applications [1–8]. In the most researches, MOFs were synthesized as homometallic MOFs: one kind of metal ion or cluster along with one type of organic linker [9, 10]. Lately, MOFs have been constructed with the further varieties as the use of at least two kinds of metals named "Mixed-Metal MOFs" term (MM′-MOFs) [11–16]. As well, the "Mixed-Ligand MOFs" term is employed as the mixed-component MOFs (MC-MOFs) that has been explained in another chapter [17–19]. There are some limitations in the exchange of MOFs components that are often complicated so it is concluded that the metal exchange control and distribution are substantial difficult. But it is interesting that during transmetalation in MOFs and creation of the

Ali Morsali and Kayhaneh Berijani. Metal-Organic Frameworks with Heterogeneous Structures, (71–104) © 2021 Scrivener Publishing LLC

isostructure MOFs, structural integrity can be remained [20]. In the production of complexity in MOFs, the varied methods exist but the common route is the synthesis of MM′-MOFs by mixing metals. In the following, we reached a deep understanding with further details about MM′-MOFs. In 2011, Chen and co-workers presented MM′-MOFs due to the adding of the dissimilar metal to anywhere in MOFs framework with any method (with or without) interaction [21]; The same metal with different oxidation states, with the following classification: (i) adding to nodes [22], (ii) immobilization [23–25], (iii) metal nanoparticles or complexes [26], and (iv) organometallic [27].

These happenings lead to MOFs that second metal is added to the parent structure in different locations and finally the produced MM′-MOF has the different ratio of the first metal/second metal. Mixed-metal MOFs can show the new excellent properties and improved efficiency compared to parent MOFs analogues with single metal [28, 29]. Actually, several kinds of SBUs in MM′-MOFs are effective factors in the structural complexity to generate new abilities in MOFs. Schematic probable procedures of the MM′-MOFs preparation have been shown in Figure 5.1.

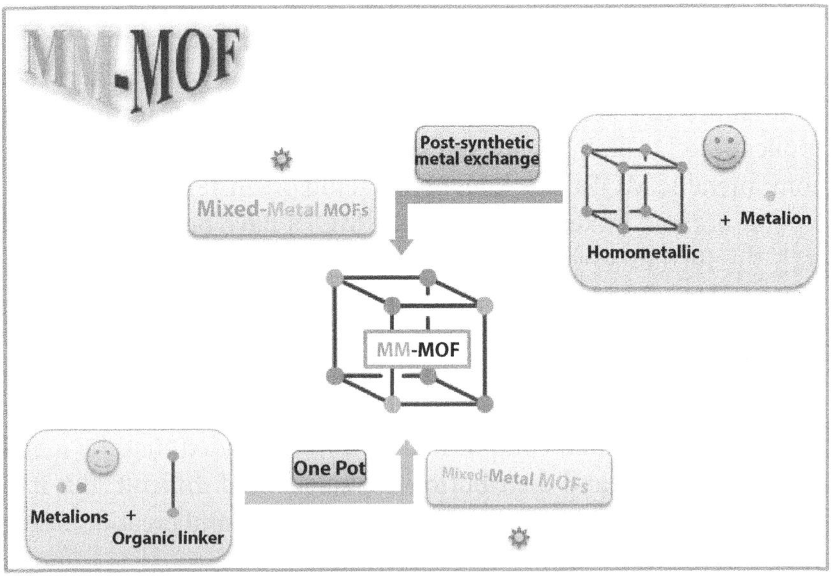

Figure 5.1 Two synthesis strategies to create MM′-MOFs.

This type of compounds is synthesized with an additional complexity and the diverse metals by the different available procedures that have been mentioned in the following:

One-pot or direct synthesis is a successful strategy to synthesize of MM′-MOFs by using the mixing of the various initial metal precursors [30–35]. An advantage of this route is the low numbers of stages. But unfortunately, the control of the metals distribution or scrambling is not possible. To solve this problem, the design of the organic linker structure needs the tuning of the various coordination sites to coordinate toward the first metal/second metal. In 2013, Baudron and co-workers reported the synthesis of the same coordination networks with the different metals by using the linkers based on dipyrrin and the first metal (M) as Cu(II), Zn(II), and Pd(II) with second metal cation (M′) like Cd(II). The used one-pot procedure is performed through the self-assembly method and the obtained results show an efficient advancement, so it can be considered as an easier method than the others. Using metalloligands is another synthetic method to generate MM′-MOFs (Figure 5.2) [36]. In fact, to synthesize this kind of MOFs, both sequential and one-pot methods have been studied. For the stepwise route, the differentiation between the two coordination sites was exploited for the preparation of discrete complexes. Acting as metallatectons, these constructed building blocks can lead to the synthesis of MM′-MOFs

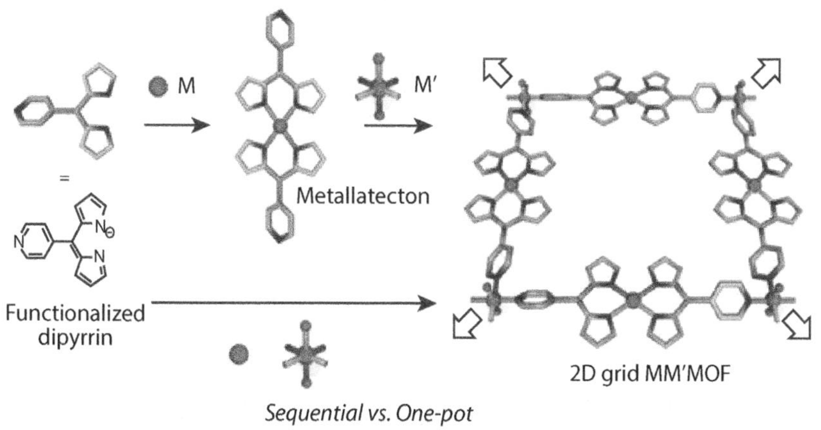

Figure 5.2 Two synthetic methods to create 2D gird MM′-MOFs [36].

upon self-assembly with the considered metal salt. In the obtained rhombic grid-type structures, four consecutive metallatectons are bridged by considered cations with chloride anions occupying apical positions. With respect to the kind of the existent metal center, MOFs can have different capabilities as photoactive, catalyst, and sensor. Elemental analysis, XPS, and EDX are the basic and routine techniques to investigate the produced mixed-metal MOFs.

Apart from the mentioned abilities, one of the most important advantages of this method is the creation of the pores with tunable sizes. Before explaining this method as a direct synthesis, it can be said that some limitations are available (more than one kind of metal in the second metal complexes). Sometimes, the used metalloligands can be chiral that lead to the chiralization of MOF [37]. In 2015, Chen *et al.* reported a Zr-MOF with two kinds of metallolinkers and two different metals, Ru and Pt (Figure 5.3) [38]. The considered

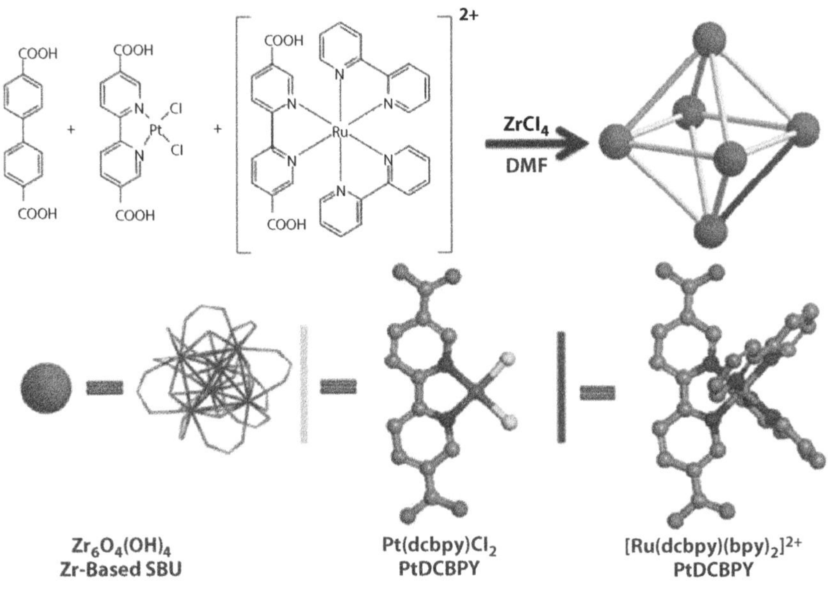

Figure 5.3 The incorporated BPDC, RuDCBPY, and PtDCBPY into Zr-MOF (UiO-67) through the solvothermal method. Red balls: SBUs. Gray rod: BPDC. Purple rod: RuDCBPY. Yellow rod: PtDCBPY. This figure is copied with the permission of the mentioned reference [38].

Zr-MOF was synthesized with octahedral cavity through solvothermal method and characterized by different techniques.

The integrity of Ru and Pt species within MOFs was shown by several techniques such as XRD, BET, TGA, SEM, TEM, HAADF-STEM, EDX, DRUS, and XPS that some of them have been demonstrated in Figure 5.4. In this work, samples $Pt_{5.4\%}$@UIO-67 and $Ru_{0.86\%}$-$Pt_{5.6\%}$@UIO-67 were investigated. As shown in Figure 5.4a, PXRD patterns indicated the retention of the parent framework after structural modification. IR of two samples Ru-Pt@UIO-67 and Pt@UIO-67 were investigated (Figure 5.4b). Considerable agreements were seen between core@shell samples and UIO-67 with minor differences in band position and shape. Small differences showed little perturbation of UIO-67 upon introduction of Ru and Pt complexes. TGA analysis of core@shell samples was also investigated under air atmosphere (Figure 5.4c).

Such as UIO-67, MOFs exhibited two decomposition steps. The first negligible mass loss is probably due to volatilization of DMF.

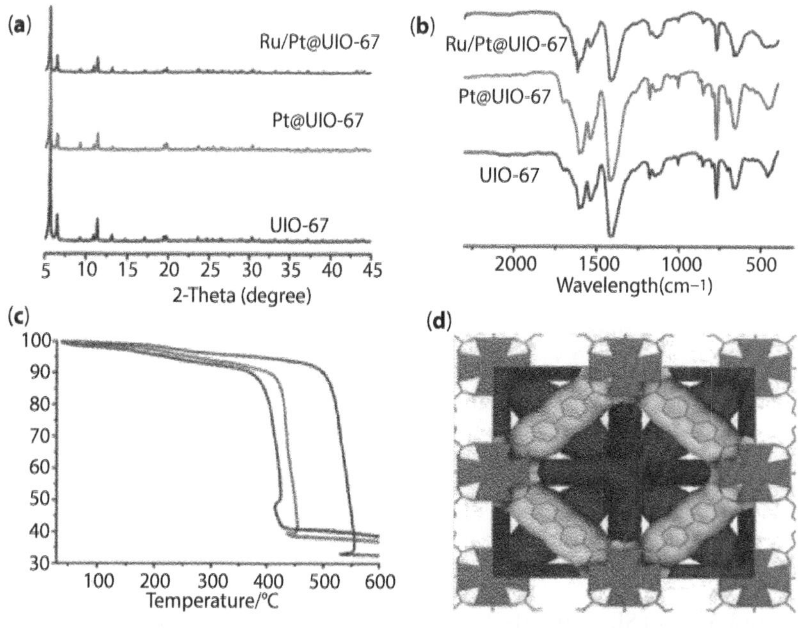

Figure 5.4 (a) Powder x-ray diffraction patterns, (b) IR spectra, and (c) TGA analysis of UIO-67_{black}, Pt@UIO-67_{red}, and Ru-Pt@UIO-67_{blue}. (d) Connolly-surface-filling model demonstrates mother MOFs of UIO-67 (pore size: 8–10 Å) [38].

Decomposition of MOFs was started at >400°C that finally a combination of ZrO_2 and considerable metal oxide phases were generated. The results showed suitable thermal stability of the obtained MOFs. SEM images showed an octahedral or semi octahedral morphology for the samples with a broad range of particle sizes, 800–1,500 nm

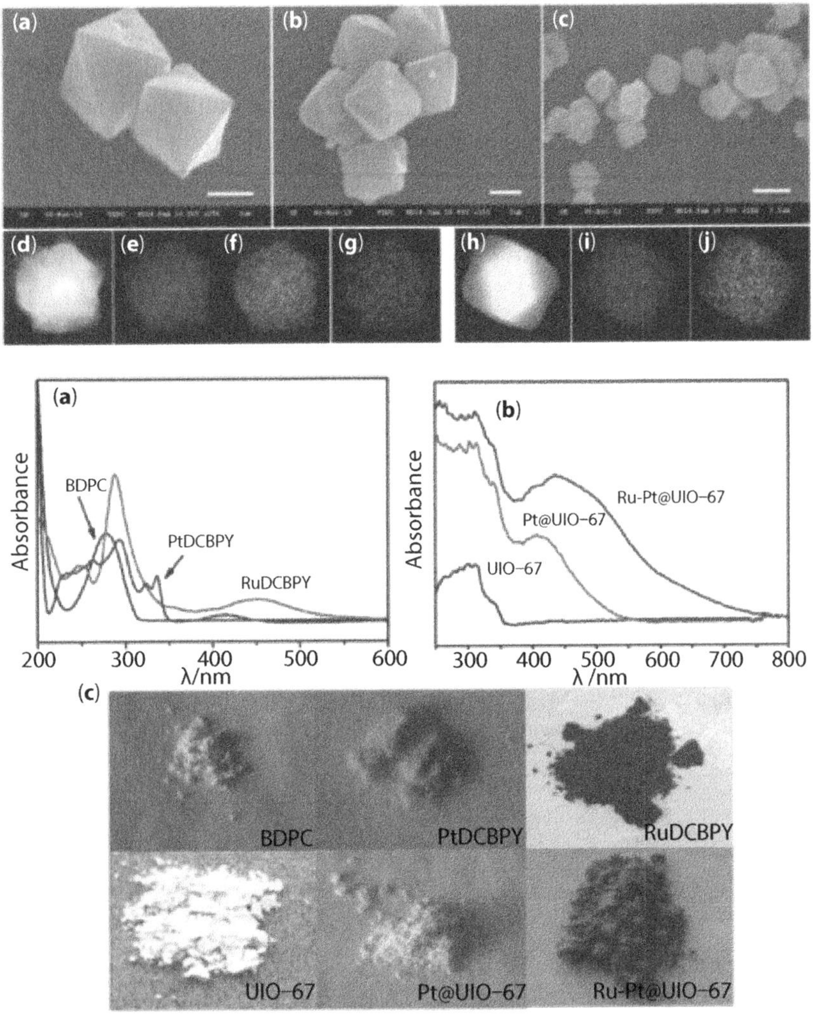

Figure 5.5 (a–c) SEM and (d–j) HAADF-STEM images of the considered samples (top). UV-vis of bpdc, RuDCBPY (aqueous solution), and PtDCBPY (MeOH). (a, b) Diffuse reflectance of the considered samples (middle). (c) Powdery samples photographs under natural light (bottom) [38].

(Figures 5.5a–c). HAADF-STEM and EDX-mappings were used to show the presence of Ru and Pt in MOFs (Figures 5.5d–j).

Uniformly distributed metal elements inside the considered samples were confirmed by these techniques: encapsulated Ru and Pt by MOFs. Diffuse reflectance UV-vis spectroscopy was another technique to investigate the samples (Figure 5.5, bottom). The absorption of Ru-Pt@UIO-67 at ~450 nm, resembling that of RuDCBPY in water, can be related to MLCT of $Ru(d\pi^6) \rightarrow Ru(d\pi^5)bpy(\pi^*)$. Also, Pt@UIO-67 demonstrates absorption peak at ~410 nm corresponding to absorption of MLCT $[Pt(d\pi) \rightarrow dcbpy(\pi^*)]$ for PtDCBPY in MeOH. This absorption peak was observed in Ru-Pt@UIO-67, too. Sometimes, the metalloligands were used in the construction of infinite frameworks due to their benefits: simple assembly of the heterometallic systems. So, di-component Ru-Pt@UIO-67 MOF assembly permits a simple arrangement of the photosensitizer and reduction heterogeneous catalyst with close spatial proximity, promotes the electron transfer between them, and therefore leads to high hydrogen evolution activity in aqueous solution upon visible light.

In 2005, Kitagawa *et al.* reported the control of crystal framework by using the metalloligand (Figure 5.6) [39]. In this work, coordination polymers were constructed by a novel metalloligand that had two different coordination groups (pyridine and carboxylate) and Cu-metalloligand. Co, Cu, Zn, and Fe were the metals of the formed coordination polymers. It was interesting that these structures

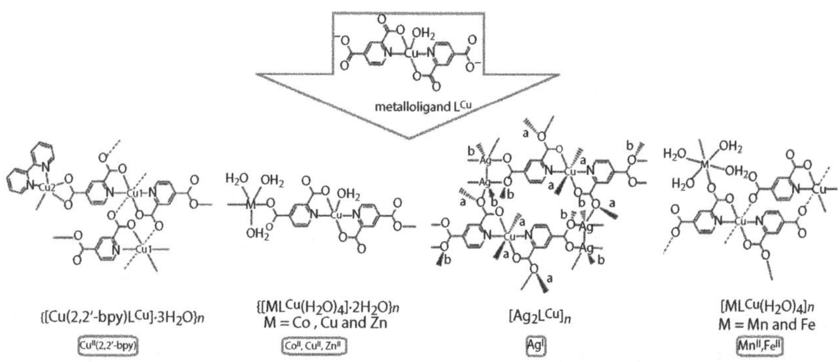

Figure 5.6 The formation of coordination polymers from the corresponding metals [4 to 8]/$[Cu(2,2'\text{-bpy})]^{2+}$[9]/[10] and metalloligand $[L^{Cu}]$ [39].

had magnetic features. Using the porphyrin ligands as metalloligands is an ideal and suitable example to generate the multivariate MM′-MOFs [MTV-MOFs], too [40].

More than one isostructural mixed-metal-organic framework (M′MOFs) have reported that they were synthesized by using both the organic ligand and metalloligand with the potential application in the enantioselective separation as chiral alcohols with high enantiomeric excess. Surely, with employing two kinds of metallic ligands, a difference in the generated frameworks is observed that is still challenging. Post-synthetic cation-exchange procedure is a useful strategy to create MOFs that are not synthesized with direct approach. For the first time, the post-synthetic metal exchange in ZIF as a stable framework in $Mn(acac)_2$ solution was investigated [41]. In this route, transmetalation happens by the metal ions replacement that leads to the construction of isostructure MOFs. This method is done through the soaking of MOF in the concentrated solution of different metal ion with parent MOF metal. This process can be proceeded for several days under the suitable conditions (temperature, solvent, and concentration of cation solution). Metal ions in the framework can affect its stabilities. When this method is used that the considered mixed-metal MOF cannot be produced by *de novo* method. The replacement of metal in the nodes, the insertion of new metal to MOF's structure and epitaxially growth are parts of the post-synthetic strategy. So, in the following, these methods have been discussed [42].

The other case of the post-synthetic cation-exchange is epitaxial growth on MOF. The synthesis of PCP-on-PCP (Porous Coordination Polymers) or MOF-on-MOF by heteroepitaxial growth can lead to the creation of heterostructures. In 2012, Wöll *et al.* published a review about epitaxially grown MOFs [43]. This method has been much utilized to generate homo-systems; so, it is logical that the heterocrystals of MOFs by heteroepitaxial growth attracts attention. As well, this strategy is used to prepare MOFs multilayers and even MOF_{core}-MOF_{shell}. Two kinds of mixed components are observed in the coordination polymers based on segregations or distributions. The first state is related to the production of the isomorphous structures along with the distributed tunable heterometals, randomly or statistically. Core-shell structure is the second state that is constructed by the epitaxial growth but chemical differences are observed in the sub-domains of crystals. Some of the prepared hetero-Ln-MOF single crystals (core-shell or striped) were reported through this route with anisotropic behavior (Figure 5.7) [44].

These compounds showed the luminescence property and even they can be used in the optoelectronic devices. As mentioned, hybrid MOF-on-MOF with different metals can be produced by heteroepitaxial growth in the liquid phase such as the growth of $[Zn_2(ndc)_2(dabco)]_n$ on $[Cu_2(ndc)_2(dabco)]_n$ that was reported by Furukawa *et al.* [45]. The happened deposition of MOF on MOF was characterized by surface plasmon resonance (SPR) that is a powerful spectroscopic method for detecting MOFs interactions during the immobilization of the MOF on another MOF. It is interesting that the heteroepitaxial growth approach can be used in the synthesis of MOF-on-MOF with the different ligands, like $Cu_2(BPYDC)_2$-on-$Cu_2(BPDC)_2$ [BPDC (biphenyl-4,4'-dicarboxylate) and BPYDC (2,2′-bipyridine-5,5′-dicarboxylate)] that its structure was corroborated by using XRD and recently it was presented by Takahashi *et al.* [46]. They reported the multiple-layered MOFs films that their deposition happened through liquid-phase epitaxy (LPE). In fact, this method as a solution method is utilized to grow crystal layers. The metal elimination-addition [47, 48], metal exchange [49], and metal insertion [50, 51] are three methods to produce MM′-MOFs.

The metal addition to MOFs can happen due to the metal elimination of MOFs frameworks (metals of nodes or the used

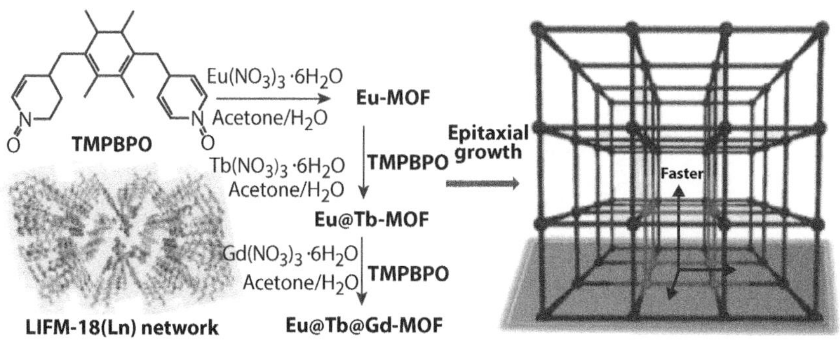

Figure 5.7 The epitaxial growth of the trimetallic and bimetallic hetero-(Ln)-MOFs from the crystal seed with one type of metal [44].

metalloligands in the frameworks) that several vacancies can be created (Figure 5.8) [52]. Also, this process can be observed in mixed-ligands MOFs. During the stirring of MOFs with one kind of metal in the second metal solution, metal exchange takes place that the content of replacing can be controlled with the effective synthetic factors. Interestingly, core-shell mixed-metal MOFs can be synthesized through this strategy. As well, metal insertion can be employed to produce mixed-metal MOFs. Insertion of different metal agents to parent MOF metal ions is another appropriate route to create MM′-MOFs such as (i) insertion of metal into open ligand site, (ii) the inclusion of metal cation to the pores of anionic MOF, and (iii) metal NPs.

For example, several NP@MOF were synthesized such as Au NPs, Cu NPs, Ni/Pd, and Pd/Au with different MOFs like MOF-5, CPL-1, MIL-53(Al), and MIL-101. Among their applications, we can mention the reduction of ketones [53], H_2 generation from hydrous hydrazine [54], CO oxidation [55], and synthesis of secondary arylamines [56].

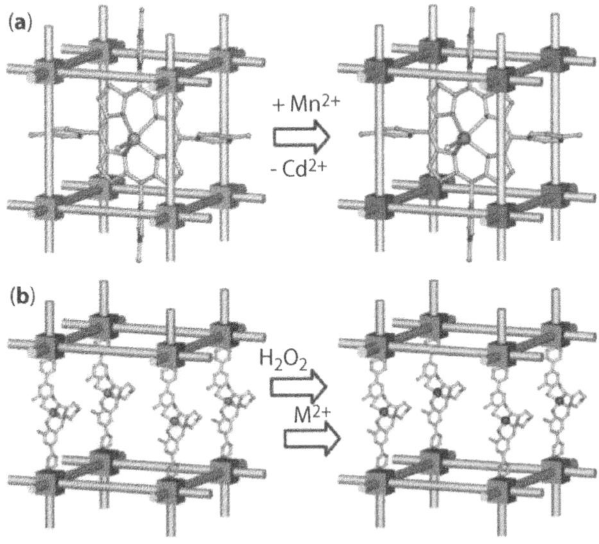

Figure 5.8 Transmetalation in porph@MOM-10 (a) and MSO-MOF (b). [Panel (b): metalation after demetalation of MOF] [52].

However, the characterization of this kind of materials is an important topic and challenging. In this section, the explanations were reported about the prepared MM′-MOFs and compared to their homometallic analogs. As examples, this kind of mixed metal matrixes as mixed-metal crystalline porous materials can have CO_2 and H_2 gas adsorption ability. The capacity of these MOFs increases during the construction methods. MM′-MOFs are applied as heterogeneous catalysts, too. Hence, the mixing of multiple metals in MOF framework can enhance the modified features. In fact, the occupied position of the different second metal is an important point in MM′-MOFs applications. They can act as heterogeneous catalyst in the different catalytic reactions. With study of this subject, we tried for the expansion of necessary points and evidences in mixed-metal MOFs. It is concluded that the metal heterogeneity in the framework can be useful for the integration of different properties and the complexity of these materials as a new line that is still a serious, considerable and challenging topic.

5.2 Multiple SBUs

Typically, MOFs are synthesized with metallic units and organic linkers. The inorganic and organic sections control MOFs topologies, features, and interactions with adsorbates, and totally, they affect MOFs applications. To design and construction of the complicated MOFs, there are different ideas. One of them is the manipulation of SBU nature as a basic organizer in the creation of MOFs. A type of SBUs gives the new properties to MOFs crystalline structures like stability, rigidity, orientation of structure, high porosity, and the other unique chemical treatments. One of the important agents in the appearance of complexity in MOFs is a type of SBU. The vast range of MOFs was constructed by single metallic nodes as metal ions or clusters (one kind of SBU). MOFs with two distinct SBUs were also synthesized that some of them have been mentioned: the construction of two unique robust MOFs was reported with distinct metallic SBUs units as Gd-cluster and Cu-cluster. This method is one of the strategies to produce of MOFs with the multiple SBUs that different metals are used in the creation of MOFs. These MOFs with $[Cu_{12}I_{12}]$ + $[Gd_3]$ and $[Cu_3I_2]$ + $[Gd_4]$ cluster SBUs can be heterogeneous catalysts with highly porous, excellent catalytic activity and high stability (Figure 5.9). Certainly $[Cu_{12}I_{12}]$ and $[Cu_3I_2]$ are the active catalytic centers. These catalysts were used in the carboxylation reactions of various terminal alkynes. Importantly, the mild conditions were employed without any co-catalyst/additive. MOFs with cluster-based were also prepared that they can accelerate chemical conversion of CO_2 through C–C bond formation which it is very valuable [57].

At first, phenylacetylene (1a) was selected as the model substrate for the optimization study of the carboxylation with CO_2. The optimization of the carboxylation of phenylacetylene and carboxylation of terminal alkynes with CO_2 data have been shown in Tables 5.1 and 5.2.

A MOF was designed with chemical formula $\{[H_2N(Me)_2][Zn_2(L)(H_2O)]\cdot DMF\cdot H_2O\}n$ with a (4,6)-connected 3D net and triclinic space group that has two kinds of binuclear clusters ($[Zn_2(\mu_2\text{-}COO)_2(\mu_1\text{-}COO)_4]$, $[Zn_2(\mu_2\text{-}COO)_4]$) with luminescent properties (Figure 5.10) [58]. The construction of a MOF was also

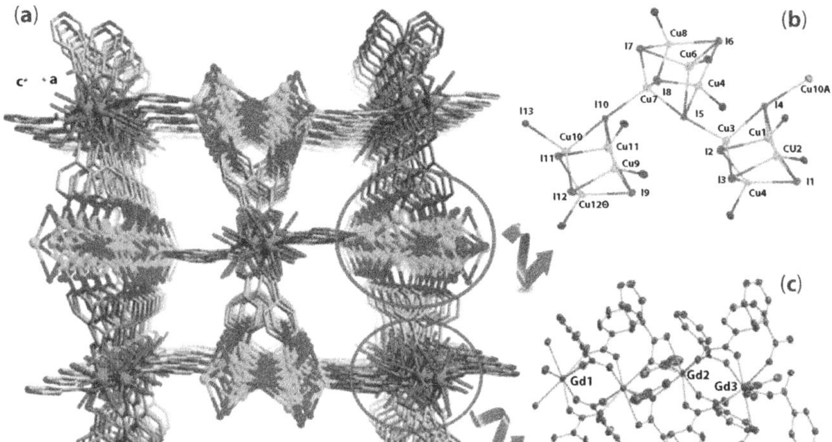

Figure 5.9 (a) 3D structure of $[Gd_3Cu_{12}I_{12}(IN)_9(DMF)_4]_n$.nDMF. **(structure I)** (b) $[Cu_{12}I_{12}]$ cluster. (c) Tri-nuclear $[Gd_3(IN)_9(DMF)_4]$ units (top). (a) 3D structure of $[Gd_4Cu_4I_3(CO_3)_2(IN)_9(HIN)_{0.5}(DMF)(H_2O)]_n$.nDMF.nH$_2$O. **(structure I)** (b) $[Cu_3I_2]$ cluster. (c) $Gd_4(CO_3)_2$ chains (bottom) [57].

Table 5.1 Optimization of the carboxylation of phenylacetylene[a].

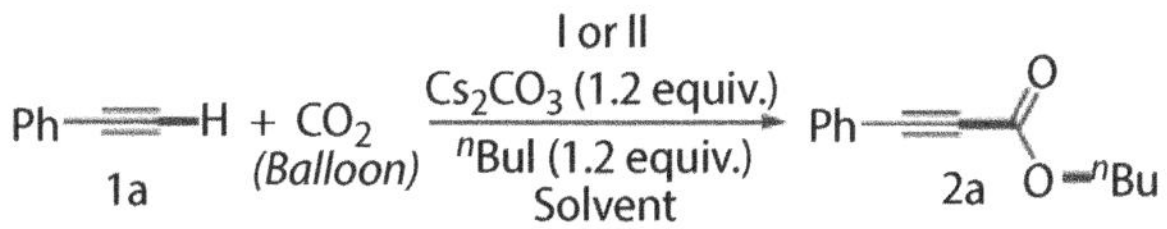

Entry	Catalyst (mol% of Cu)	Solvent	T(°C)	t(h)	Yield[b] (%)
1	–	DMF	80	12	9[c], 4[d]
2	Cat-I (7)	DMF	80	12	47
3	Cat-I (7)	DMF	100	24	17
4	Cat-I (6)	EC	80	12	61
5	**Cat-I (6)**	**EC**	**80**	**4**	**74**
6	Cat-II (3)	DMF	80	12	34
7	Cat-II (3)	DMF	100	24	28
8	Cat-II (3)	EC	80	12	70
9	**Cat-II (4)**	**EC**	**80**	**4**	**80**

[a]Reaction conditions: phenylacetylene (51 mg, 0.5 mmol), Cs_2CO_3 (196 mg,0.6 mmol), n-Bul (110 mg, 0.6 mmol), DMF or ethylene carbonate (EC,3 ml), CO_2 (99.999%, balloon), indicated amount of catalyst, 12 h. [b]The yields were determined by GC with biphenyl as the internal standard. [c]Without a catalyst. [d]Without CO_2.

investigated through interpenetration of cationic and anionic nets ($[Zn_7(L_1)_3(H_2O)_7]_n.[Zn_5(L_8)_3(H_2O)_5]_n$ = Zn-MOF) with two kinds of ligands, that L_1 is tetrazolate and L_8 is tetracarboxylic acid. In the frameworks of final MOF, there were two types of Zn-containing SBUs (square pyramidal and triangular-paddlewheel).

So, one method to form MOFs with distinct SBUs is the presence of the different kinds of SBUs from the point of geometrically even with the same metals.

In the above MOF, with two different linkers, the metal exchange with $CoCl_2$, $CuCl_2$, and $NiCl_2$ was studied, too. The Cu-Zn-MOF was generated with 87% exchange but about two other metals it did not happen due to ion diameter and the ability to coordination [59]. In the last years, some examples have been presented like

Table 5.2 Carboxylation of terminal alkynes with CO_2[a].

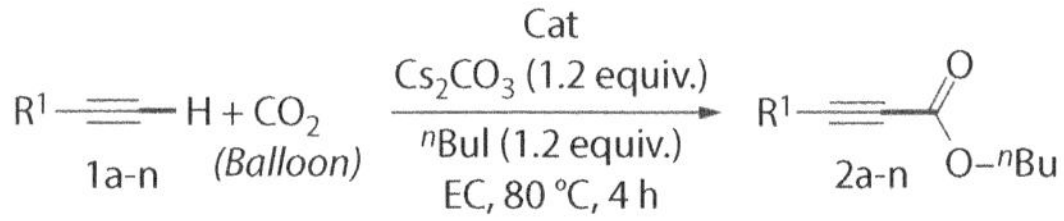

Entry	$R^1C{\equiv}CH$	Product	Yield[b] (%)	
			A	B
1	 **1a**	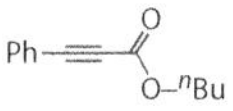 **2a**	80	86
2	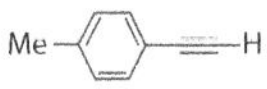 **1b**	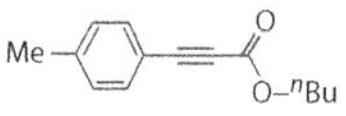 **2b**	70	77
3	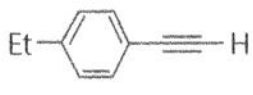 **1c**	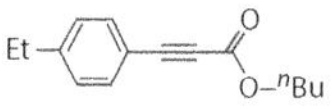 **2c**	73	85
4	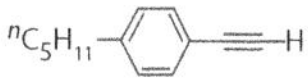 **1d**	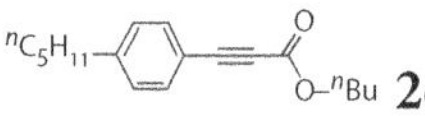 **2d**	81	82
5	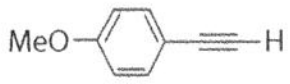 **1e**	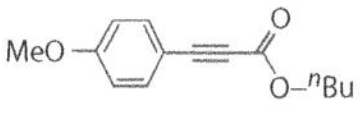 **2e**	65	74
6	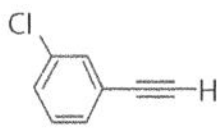 **1f**	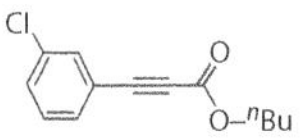 **2f**	72	73
7	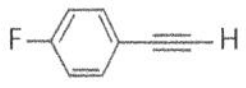 **1g**	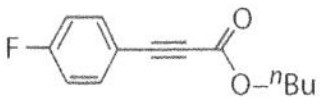 **2g**	67	70
8	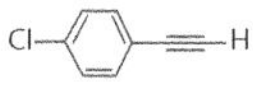 **1h**	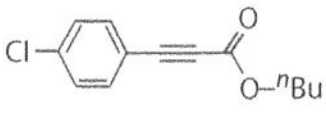 **2h**	60	74
9	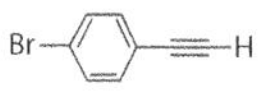 **1i**	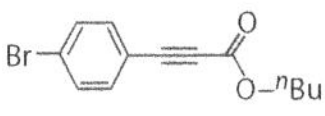 **2i**	56	69
10	 **1j**	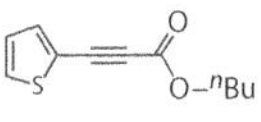 **2j**	76	80
11	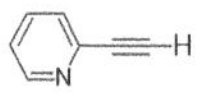 **1k**	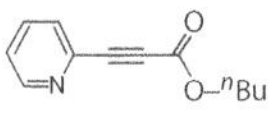 **2k**	66	77
12	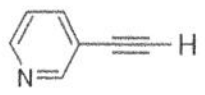 **1l**	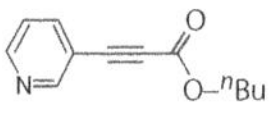 **2l**	58	65

(Continued)

Table 5.2 Carboxylation of terminal alkynes with CO_2[a]. (*Continued*)

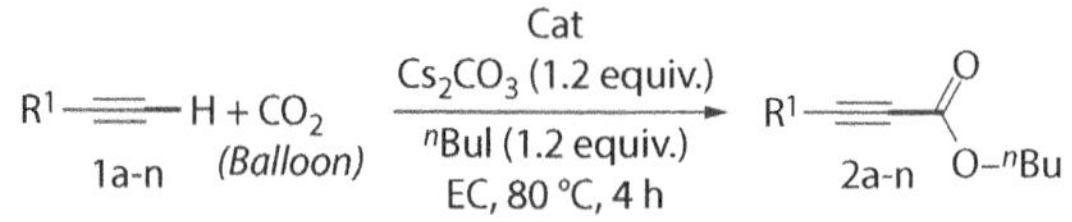

Entry	$R^1C{=}CH$	Product	Yield[b] (%)	
			A	B
13	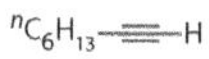 **1m**	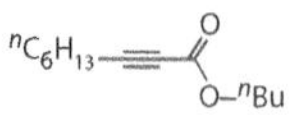 **2m**	71	75
14[c]	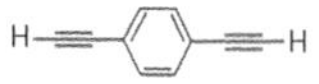 **1n**	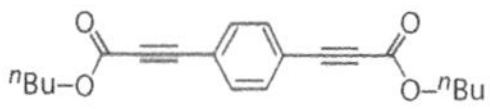 **2n**	52	60

[a]Reaction conditions: terminal alkyne (1.0 mmol), catalyst I (6 mol%, method A) or catalyst II (4 mol%, method B), Cs_2CO_3 (0.391 g, 1.2 mmol), *n*-Bul (0.221 g, 1.2 mmol) and ethylene carbonate (3 ml), CO_2 (99.999%, balloon), 80°C, 4 h. [b]Isolated yield. [c]I (12 mol%) or I I (8 mol%), Cs_2CO_3 (2.4 mmol), *n*-Bul (2.4 mmol), EC (5 ml) were used.

the mixed-coordination metal-imidazolate framework with Zn^{2+}-tetrahedral and In^{3+}-octahedral [60], the prepared MOF with tritopic linkers and dinuclear metal SBUs with the same metal and different coordination geometries, Zn-square and Zn-tetrahedral [61],

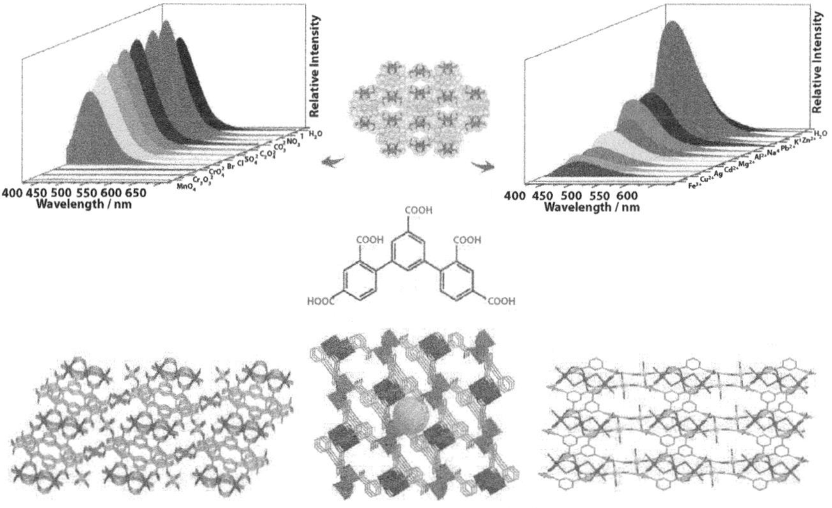

Figure 5.10 The solid-state luminescent properties of MOFs with diverse SBUs [58].

and PCN-515 crystals with three different kinds of Zn-SBUs geometrically [62]. Recently, the various MOFs based on multiple SBUs were reported to form triangular/square, square/tetrahedral, square/trigonal prismatic, square/octahedral, tetrahedral/trigonal prismatic, square pyramidal/octahedral SBUs, and triangular/square/hexatopic linker.

These numbers of SBUs in framework can induce heterogeneity to net and reveal intriguing properties than single SBU [63]. Yaghi and co-workers reported a MOF that was synthesized during the solvothermal reaction with one kind of organic linker, 4-pyrazolecarboxylic acid (H_2PyC), and three different SBUs that were Cu-based triangular, Zn-based octahedral, and square pyramidal SBUs [64]. From the combination of these compounds together, four types of cages are built. These complex cages can provide environment to incorporate catalytical sites, different functional groups, and guest

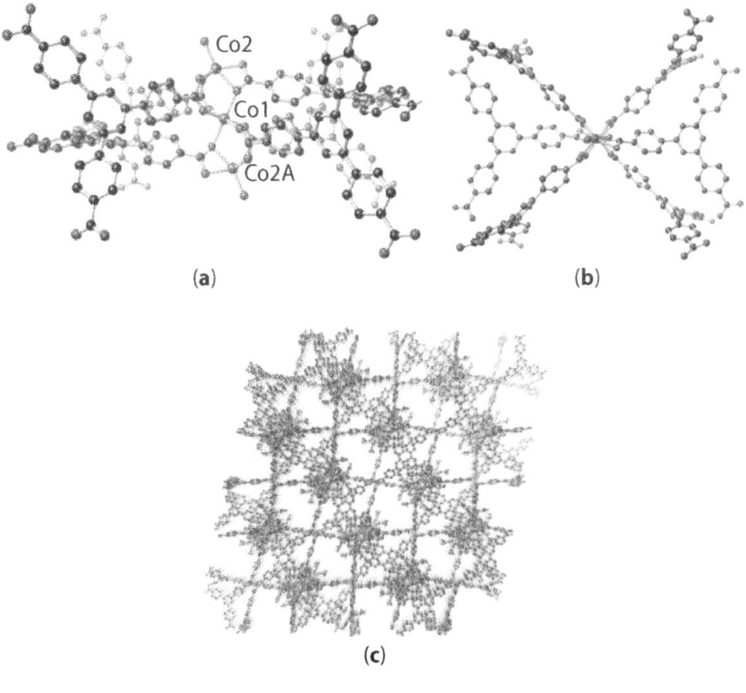

Figure 5.11 (a and b) MOF 1 structure showing Co_3 SBU supported by Co1 and Co2 viewing from two directions. (c) Three-dimensional network viewing along the [010] direction [75].

molecules. The "heterogeneity within order" term can be used for this mesoporous MOF [65, 66]. So, the induced heterogeneity to framework can be due to the distinct geometries in the inorganic SBUs that lead to the varied pores [67–72]. Sometimes, in the organic linkers, the orientational disorders exist that they generate variations.

Table 5.3 Crystallographic data and refinement parameters of the considered MOF.

Parameter	Value
Molecular formula	$C_{172}H_{112}C0_{114}N_6O_{64}$
Formula weight	3934.90
Crystal system	Monoclinic
Space group	$P2_1/n$
a(Å)	27.295(3)
b(Å)	36.845(4)
c (Å)	29.289(3)
β(°)	108.940(2)
V (Å^3)	27,861(5)
Z	2
P_{calc} (g cm^{-3})	0.469
F(000)	3990
μ(mm^{-1})	0.347
Total reflections	483,464
Unique reflections	49,039
Observed reflections	21,553
R_{int}	0.1868
Variables	1138
$R_1{}^a$	0.1225

Table 5.4 Various applications of mixed-metal MOFs.

Entry	Mixed metal MOFs	Application	Ref.
1	CPM-200 (Sc^{3+}/Mg^{2+}, Mg^{2+}/Fe^{3+}, Mg^{2+}/Ga^{3+})	CO_2,H_2 adsorption	[77]
2	MOF-5 (Zn^{2+}/Co^{2+})	H_2,CH_4,CO_2 adsorption	[78]
3	MIL-101 (Cr^{3+}/Mg^{2+})	CO_2 adsorption, CO_2/N_2 selectivity	[79]
4	MOF-74 (Zn^{2+}/Co^{2+})	H_2,CH_4,CO_2 uptakes	[80]
5	MOF-74 (Co^{2+}/Ni^{2+})	H_2 adsorbent	[81]
6	COMOC-2 (Al^{3+}/V^{4+})	CO_2 adsorption	[82]
7	Mg-MOF-74 (Mg^{2+}/Co^{2+} or Ni^{2+})	CO_2 adsorption	[83]
8	HKUST-1 (Cu^{2+}/Zn^{2+})	HD adsorption and desorption	[84]
9	Mg-MOF-74 (Cd^{2+}/Mg^{2+})	CO_2 adsorption	[85]
10	MIL-53 (Cr^{3+}/Fe^{3+})	CO_2 adsorption	[86]
11	MIL-53 (Cr^{3+}/V^{3+})	CO_2 adsorption	[87]
12	CTOF-1 (Co^{2+}/Ti^{4+}) CTOF-2	H_2,CH_4,CO_2 adsorption	[88]
13	UiO-66 (Zr^{4+}/Ti^{4+})	CO_2 adsorption	[89]
14	Ni-ITHD (Ni^{2+}/Zn^{2+}, Ni^{2+}/Co^{2+})	CO_2 uptake	[90]
15	PCN-922 (Zn^{2+}/Cu^{2+})	H_2,CH_4,CO_2 adsorption	[91]
16	CPM-18 to 21 (In^{3+}/Nd^{3+}, Sm^{3+}, Pr^{3+}, Mn^{2+}, Co^{2+}, Cu^{2+}, Mg^{2+})	H_2, CO_2 adsorption	[92]

(Continued)

Table 5.4 Various applications of mixed-metal MOFs. (*Continued*)

Entry	Mixed metal MOFs	Application	Ref.
17	$Mn_3[(Mn_4Cl)_3(BTT)_8$-$(CH_3OH)_{10}]_2$ (Mn^{2+}/Li^+, Cu^+, Fe^{2+}, Co^{2+}, Ni^{2+}, Cu^{2+}, Zn^{2+})	H_2 storage	[93]
18	CMP-15 (In^{3+}/Co^{2+}, Mg^{2+}, Mn^{2+}, Ni^{2+}, Cd^{2+})	H_2, CO_2 adsorption	[92]
19	ZIF-8 (Co^{2+}/Zn^{2+})	H_2, CO_2 uptakes	[94]
20	$Cd_3[(Cd_4Cl)_3(BTT)_8]_2$ (Cd^{2+}/Co^{2+}, Cd^{2+}/Ni^{2+})	H_2 adsorption	[95]
21	MIL-53 (Al^{3+}/Cr^{3+})	CO_2 adsorption	[96]
22	ZTOF-1 (Zn^{2+}/Ti^{4+})	H_2, CO_2 adsorption	[97]
23	UiO-66 (Zr^{4+}/Ti^{4+})	CO_2 separation	[98]
24	isol (Mn^{2+}/Cu^{2+})	CO_2/CH_4 separation, MeOH/CH_3CN and EtOH vapor separation	[99]
25	HKUST-1 (Cu^{2+}/Li^+, Na^+, K^+)	CO_2 adsorption	[100]
26	MIL-53 (Al^{3+}/V^{3+})	Glycerol condensation with acetone	[101]
27	MOF-74 (Cu^{2+}/Co^{2+})	Styrene oxidation	[102]
28	MIL-100 (Sc^{3+}/Fe^{3+})	Alcohol oxidation, tandem C–C bond formation	[103]
29	MOF-74 (Ni^{2+}/Co^{2+})	Oxidation of cyclohexene	[104]
30	$\{[CuM(pdc)_2(H_2O)_X]\cdot YH_2O\}_n$ (Cu^{2+}/Mg^{2+}, Cu^{2+}/Ca^{2+}, Cu^{2+}/Sr^{2+}, Cu^{2+}/Ba^{2+}) (X = 0, 3, 4, 5, Y = 0–2)	Epoxidation of olefins	[105]

(*Continued*)

Table 5.4 Various applications of mixed-metal MOFs. (*Continued*)

Entry	Mixed metal MOFs	Application	Ref.
31	UiO-66 ($Zr^{4+}/Ce^{3+,4+}$)	Decomposition of methanol into CO_2	[106]
32	MIL-101 ($Cr^{3+}/Ce^{3+,4+}$)	H_2 production from ammonia borane	[107]
33	ZIF-8 (Zn^{2+}/Cu^{2+})	Cycloaddition of organic azides	[108]
34	[CoNi(μ_3-tp)$_2$(μ_2-pyz)$_2$] (Co^{2+}/Ni^{2+})	Dye removal	[109]
35	MFM-300(Ga)$_2$ (Ga^{3+}/Fe^{3+})	Acetylation of benzaldehyde	[110]
36	{[Zn_2(L)(H_2O)$_2$].(5DMF).(4H_2O)} (Zn^{2+}/Cu^{2+})	Knoevenagel condensation	[111]
37	Porph@MOM-10 (Cd^{2+}/Mn^{2+}, Cd^{2+}/Cu^{2+})	Epoxidation oftrans-stilbene	[112]
38	[$In_xGa_{1-x}(O_2C_2H_4)_{0.5}$(hfipbb)] ($Ga^{3+}/In^{3+}$)	A^3 Strecker reaction	[113]
39	ZIF-8 (Zn^{2+}/Ni^{2+})	Electroreduction of CO_2	[114]
40	NH_2-UiO-66 (Zr^{4+}/Ti^{4+})	CO_2 reduction and hydrogen evolution	[115]
41	UiO-66 (Zr^{4+}/Ti^{4+})	CO_2 reduction to HCOOH	[116]
42	UiO-67 (Zr^{4+}/Ti^{4+})	Degradation of methylene blue	[117]
43	UiO-66 (Zr^{4+}/Ti^{4+})	PCVG to reduce Se^{6+}	[118]
44	NDC-MOFs (Zr^{4+}/Ti^{4+})	Cascade MPV reduction and Etherification	[119]

(*Continued*)

Table 5.4 Various applications of mixed-metal MOFs. (*Continued*)

Entry	Mixed metal MOFs	Application	Ref.
45	MIL-101 (Cr^{3+}/Fe^{3+})	Dye degradation	[120]
46	MMPF-5 (Cd^{2+}/Co^{2+})	Epoxidation oftrans-stilbene	[121]

H_2Pdc, pyridine-2,5-dicarboxylic acid; tp, terephthalic acid; pyz, pyrazine; MFM-300$(Ga)_2$. ($[Ga_2(OH)_2(L)]$; H_4L, biphenyl-3,3′,5,5′-tetracarboxylic acid); L, 2′-amino-[1,1′:3′,1″-terphenyl]-3,3″,5,5″-tetracarboxylic acid ligand; porph@MOM-10, $[Cd_6(BPT)_4Cl_4(H_2O)_4].[C_{44}H_{36}N_8CdCl]$. $[H_3O]$.[solvent]; H_2hfipbb, 4,4′-(hexafluoroisopropylidene) bis(benzoic acid); PCVG, photochemical vapor generation; NDC, 2,6-naphthalendicarboxylate; MPV, Meerwein-Ponndorf-Verley; MMPF, metalmetalloporphyrin framework; tdcmpp, tetrakis(3,5 dicarboxymethylesterphenyl)porphine.

Even, variations are obvious in SBUs structure without exiting out from inherent order [73, 74]. Analysis of these kinds of MOFs gives us important information about inorganic units as distinct SBUs, organic linkers, functions, different geometries in components, how linking of the linkers with metal sites and multiple cages. Zhang *et al.* investigated the creation of a MOF as block purple crystals with mixed-cluster SBUs. SBUs consist of a rare combination of a linear trinuclear Co_3 and two kinds of dinuclear Co_2 (Co^{2+} octahedral-tetrahedral equilibrium) with large ligand [1,3,5-tris(4-carboxyphenyl)benzene (H_3BTB)]. This synthesis was done by a simple one-pot solvothermal reaction in N,N-diethylformamide, under mild conditions and the yield of obtained product was 87%. The ability of this 3D MOF was studied in sorption CO_2 [75]. It is notable that among six BTB ligands associated with Co_3 SBU, two are functioning as the in-plane ligands propagating the structure within the (101) plane, while the other four are largely responsible for the epitaxial propagation of the structure along the [010] direction (Figures 5.11a–c). Table 5.3 shows crystallographic data and refinement parameters for considered MOF.

Using multiple SBUs is a successful strategy in the development of the reticular chemistry of MOFs framework. However, making this kind of MOFs is difficult due to the limitation of the formation condition [76]. Table 5.4 shows a full overview of MM′-MOFs for various applications such as catalytic reactions, gas adsorption and separation.

5.3 Conclusion

Mixed-metal MOFs are frameworks that contain at least two different nodes. They are relatively easily synthesized by either a one-pot synthetic strategy with a synthesis mixture containing the different sources of metals or an ion-exchange strategy post-synthetically by soaking a monometallic framework in a concentrated solution of a different metal-ion. However, their characterization is more difficult. The used characterization methods are as XAS and MRM. Computational tools at multiple scales are employed, too. In several applications, mixed-metal MOFs can be used since they can show high performance than monometallic MOFs such as catalysis in the cascade or tandem processes, luminescence, and sensing.

However, there are still many challenges for the future in this field like stability, leaching of metal, and optimized synthesis. To synthesis of MOFs with complexity, there are various cases that one of them is manipulation of SBU nature in framework that these new SBUs can give new features to MOFs.

References

1. Lee, J., Farha, O.K., Roberts, J., Scheidt, K.A., Nguyen, S.T., Hupp, J.T., Metal–organic framework materials as catalysts. *Chem. Soc. Rev.*, *38*, 5, 1450–1459, 2009.
2. Isaeva, V. and Kustov, L.M., The application of metal-organic frameworks in catalysis. *Pet. Chem.*, *50*, 3, 167–180, 2010.
3. Corma, A., García, H., Llabrés i Xamena, F.X., Engineering metal organic frameworks for heterogeneous catalysis. *Chem. Rev.*, *110*, 8, 4606–4655, 2010.
4. Farrusseng, D., Aguado, S., Pinel, C., Metal–organic frameworks: opportunities for catalysis. *Angew. Chem. Int. Ed.*, *48*, 41, 7502–7513, 2009.
5. Dhakshinamoorthy, A., Alvaro, M., Corma, A., Garcia, H., Delineating similarities and dissimilarities in the use of metal organic frameworks

and zeolites as heterogeneous catalysts for organic reactions. *Dalton Trans.*, *40*, 24, 6344–6360, 2011.
6. Yoon, M., Srirambalaji, R., Kim, K., Homochiral metal–organic frameworks for asymmetric heterogeneous catalysis. *Chem. Rev.*, *112*, 2, 1196–1231, 2011.
7. Masoomi, M.Y., Morsali, A., Dhakshinamoorthy, A., García, H., Mixed-Metal MOFs: Unique Opportunities in Metal-organic Framework Functionality and Design. *Angew. Chem. Int. Ed.*, *58*, 43, 15188–15205, 2019.
8. Mu, W., Liu, D., Yang, Q., Zhong, C., Computational study of the effect of organic linkers on natural gas upgrading in metal–organic frameworks. *Microporous Mesoporous Mater.*, *130*, 1–3, 76–82, 2010.
9. Berijani, K. and Morsali, A., Dual activity of durable chiral hydroxyl-rich MOF for asymmetric catalytic reactions. *J. Catal.*, *378*, 28–35, 2019.
10. Kökçam-Demir, Ü., Goldman, A., Esrafili, L., Gharib, M., Morsali, A., Weingart, O., Janiak, C. Coordinatively unsaturated metal sites (open metal sites) in metal–organic frameworks: design and applications. *Chem. Soc. Rev.*, *49*, 9, 2751–2798, 2020.
11. Lee, Y., Kim, S., Kang, J.K., Cohen, S.M., Photocatalytic CO_2 reduction by a mixed metal (Zr/Ti), mixed ligand metal–organic framework under visible light irradiation. *Chem. Commun.*, *51*, 26, 5735–5738, 2015.
12. Zhao, B., Chen, X.-Y., Cheng, P., Liao, D.-Z., Yan, S.-P., Jiang, Z.-H., Coordination polymers containing 1D channels as selective luminescent probes. *J. Am. Chem. Soc.*, *126*, 47, 15394–15395, 2004.
13. Guo, H., Zhu, Y., Qiu, S., Lercher, J.A., Zhang, H., Coordination Modulation Induced Synthesis of Nanoscale Eu1-xTbx-Metal-Organic Frameworks for Luminescent Thin Films. *Adv. Mater.*, *22*, 37, 4190–4192, 2010.
14. Vuong, G.-T., Pham, M.-H., Do, T.-O., Synthesis and engineering porosity of a mixed metal Fe 2 Ni MIL-88B metal–organic framework. *Dalton Trans.*, *42*, 2, 550–557, 2013.
15. Lau, C.H., Babarao, R., Hill, M.R., A route to drastic increase of CO_2 uptake in Zr metal organic framework UiO-66. *Chem. Commun.*, *49*, 35, 3634–3636, 2013.
16. Chen, B., Xiang, S., Qian, G., Metal–organic frameworks with functional pores for recognition of small molecules. *Acc. Chem. Res.*, *43*, 8, 1115–1124, 2010.
17. Bae, Y.-S., Mulfort, K.L., Frost, H., Ryan, P., Punnathanam, S., Broadbelt, L.J., Hupp, J.T., Snurr, R.Q., Separation of CO_2 from CH_4

using mixed-ligand metal–organic frameworks. *Langmuir*, *24*, 16, 8592–8598, 2008.
18. Ma, B.-Q., Mulfort, K.L., Hupp, J.T., Microporous pillared paddle-wheel frameworks based on mixed-ligand coordination of zinc ions. *Inorg. Chem.*, *44*, 14, 4912–4914, 2005.
19. Casado, F.J.M., Cañadillas-Delgado, L., Cucinotta, F., Guerrero-Martínez, A., Riesco, M.R., Marchese, L., Cheda, J.A.R., Luminescent lead (II) complexes: new three-dimensional mixed ligand MOFs. *CrystEngComm*, *14*, 8, 2660–2668, 2012.
20. Lalonde, M., Bury, W., Karagiaridi, O., Brown, Z., Hupp, J.T., Farha, O.K., Transmetalation: Routes to metal exchange within metal–organic frameworks. *J. Mater. Chem. A*, *1*, 18, 5453–5468, 2013.
21. Das, M.C., Xiang, S., Zhang, Z., Chen, B., Functional mixed metal–organic frameworks with metalloligands. *Angew. Chem. Int. Ed.*, *50*, 45, 10510–10520, 2011.
22. Noh, H., Cui, Y., Peters, A.W., Pahls, D.R., Ortuño, M.A., Vermeulen, N.A., Cramer, C.J., Gagliardi, L., Hupp, J.T., Farha, O.K., An exceptionally stable metal–organic framework supported molybdenum (VI) oxide catalyst for cyclohexene epoxidation. *J. Am. Chem. Soc.*, *138*, 44, 14720–14726, 2016.
23. Berijani, K., Morsali, A., Hupp, J.T., An Effective Strategy for Creating Asymmetric MOFs for Chirality Induction: A Chiral Zr-based MOF for Enantioselective Epoxidation. *Catal. Sci. Technol.*, *9*, 3388–3397, 2019.
24. Kitaura, R., Onoyama, G., Sakamoto, H., Matsuda, R., Noro, S. i., Kitagawa, S., Immobilization of a metallo schiff base into a microporous coordination polymer. *Angew. Chem. Int. Ed.*, *43*, 20, 2684–2687, 2004.
25. Sakamoto, H., Matsuda, R., Bureekaew, S., Tanaka, D., Kitagawa, S., A porous coordination polymer with accessible metal sites and its complementary coordination action. *Chem.–Eur. J.*, *15*, 20, 4985–4989, 2009.
26. Dhakshinamoorthy, A. and Garcia, H., Catalysis by metal nanoparticles embedded on metal–organic frameworks. *Chem. Soc. Rev.*, *41*, 15, 5262–5284, 2012.
27. Müller, M., Lebedev, O., II, Fischer, R.A., Gas-phase loading of [Zn 4 O (btb) 2](MOF-177) with organometallic CVD-precursors: inclusion compounds of the type [L n M] a@ MOF-177 and the formation of Cu and Pd nanoparticles inside MOF-177. *J. Mater. Chem.*, *18*, 43, 5274–5281, 2008.

28. Furukawa, H., Cordova, K.E., O'Keeffe, M., Yaghi, O.M., The chemistry and applications of metal-organic frameworks. *Science, 341*, 6149, 1230444, 2013.
29. Choi, K.M., Jeong, H.M., Park, J.H., Zhang, Y.-B., Kang, J.K., Yaghi, O.M., Supercapacitors of nanocrystalline metal–organic frameworks. *ACS Nano, 8*, 7, 7451–7457, 2014.
30. Caskey, S.R. and Matzger, A.J., Selective metal substitution for the preparation of heterobimetallic microporous coordination polymers. *Inorg. Chem., 47*, 18, 7942–7944, 2008.
31. Kong, X.-J., Ren, Y.-P., Long, L.-S., Huang, R.-B., Zheng, L.-S., Kurmoo, M., Influence of reaction conditions on the channel shape of 3d-4f heterometallic metal–organic framework. *CrystEngComm, 10*, 10, 1309–1314, 2008.
32. Wang, Y., Bredenkötter, B., Rieger, B., Volkmer, D., Two-dimensional metal–organic frameworks (MOFs) constructed from heterotrinuclear coordination units and 4, 4′-biphenyldicarboxylate ligands. *Dalton Trans.*, 6, 689–696, 2007.
33. He, J., Yu, J., Zhang, Y., Pan, Q., Xu, R., Synthesis, structure, and luminescent property of a heterometallic metal- organic framework constructed from rod-shaped secondary building blocks. *Inorg. Chem., 44*, 25, 9279–9282, 2005.
34. Zhang, Y., Chen, B., Fronczek, F.R., Maverick, A.W.A., Nanoporous Ag–Fe Mixed-Metal- Organic Framework Exhibiting Single-Crystal-to-Single-Crystal Transformations upon Guest Exchange. *Inorg. Chem., 47*, 11, 4433–4435, 2008.
35. Fu, Y., Xu, L., Shen, H., Yang, H., Zhang, F., Zhu, W., Fan, M., Tunable catalytic properties of multi-metal–organic frameworks for aerobic styrene oxidation. *Chem. Eng. J., 299*, 135–141, 2016.
36. Beziau, A., Baudron, S.A., Fluck, A., Hosseini, M.W., From Sequential to One-Pot Synthesis of Dipyrrin Based Grid-Type Mixed Metal–Organic Frameworks. *Inorg. Chem., 52*, 24, 14439–14448, 2013.
37. Cho, S.-H., Ma, B., Nguyen, S.T., Hupp, J.T., Albrecht-Schmitt, T.E.A., metal–organic framework material that functions as an enantioselective catalyst for olefin epoxidation. *Chem. Commun.*, 24, 2563–2565, 2006.
38. Hou, C.-C., Li, T.-T., Cao, S., Chen, Y., Fu, W.-F., Incorporation of a [Ru (dcbpy)(bpy) 2] 2+ photosensitizer and a Pt(dcbpy)Cl_2 catalyst into metal–organic frameworks for photocatalytic hydrogen evolution from aqueous solution. *J. Mater. Chem. A, 3*, 19, 10386–10394, 2015.

39. Noro, S.-i., Miyasaka, H., Kitagawa, S., Wada, T., Okubo, T., Yamashita, M., Mitani, T., Framework control by a metalloligand having multi-coordination ability: new synthetic approach for crystal structures and magnetic properties. *Inorg. Chem.*, *44*, 1, 133–146, 2005.
40. Liu, Q., Cong, H., Deng, H., Deciphering the spatial arrangement of metals and correlation to reactivity in multivariate metal–organic frameworks. *J. Am. Chem. Soc.*, *138*, 42, 13822–13825, 2016.
41. Fei, H., Cahill, J.F., Prather, K.A., Cohen, S.M., Tandem postsynthetic metal ion and ligand exchange in zeolitic imidazolate frameworks. *Inorg. Chem.*, *52*, 7, 4011–4016, 2013.
42. Deria, P., Mondloch, J.E., Karagiaridi, O., Bury, W., Hupp, J.T., Farha, O.K., Beyond post-synthesis modification: evolution of metal–organic frameworks via building block replacement. *Chem. Soc. Rev.*, *43*, 16, 5896–5912, 2014.
43. Gliemann, H. and Wöll, C., Epitaxially grown metal-organic frameworks. *Mater. Today*, *15*, 3, 110–116, 2012.
44. Pan, M., Zhu, Y.X., Wu, K., Chen, L., Hou, Y.J., Yin, S.Y., Wang, H.P., Fan, Y.N., Su, C.Y., Epitaxial Growth of Hetero-Ln-MOF Hierarchical Single Crystals for Domain-and Orientation-Controlled Multicolor Luminescence 3D Coding Capability. *Angew. Chem. Int. Ed.*, *56*, 46, 14582–14586, 2017.
45. Shekhah, O., Hirai, K., Wang, H., Uehara, H., Kondo, M., Diring, S., Zacher, D., Fischer, R.A., Sakata, O., Kitagawa, S., MOF-on-MOF heteroepitaxy: perfectly oriented $[Zn_2(ndc)_2(dabco)]_n$ grown on $[Cu_2(ndc)_2(dabco)]_n$ thin films. *Dalton Trans.*, *40*, 18, 4954–4958, 2011.
46. Ikigaki, K., Okada, K., Tokudome, Y., Toyao, T., Falcaro, P., Doonan, C.J., Takahashi, M., MOF-on-MOF: Oriented Growth of Multiple Layered Thin Films of Metal–Organic Frameworks. *Angew. Chem. Int. Ed.*, *131*, 21, 6960–6964, 2019.
47. Tu, B., Pang, Q., Wu, D., Song, Y., Weng, L., Li, Q., Ordered vacancies and their chemistry in metal–organic frameworks. *J. Am. Chem. Soc.*, *136*, 41, 14465–14471, 2014.
48. Shultz, A.M., Sarjeant, A.A., Farha, O.K., Hupp, J.T., Nguyen, S.T., Post-synthesis modification of a metal–organic framework to form metallosalen-containing MOF materials. *J. Am. Chem. Soc.*, *133*, 34, 13252–13255, 2011.
49. Song, X., Kim, T.K., Kim, H., Kim, D., Jeong, S., Moon, H.R., Lah, M.S., Post-synthetic modifications of framework metal ions in isostructural metal–organic frameworks: Core–shell heterostructures via selective transmetalations. *Chem. Mater.*, *24*, 15, 3065–3073, 2012.

50. Yang, S., Lin, X., Blake, A.J., Walker, G.S., Hubberstey, P., Champness, N.R., Schröder, M., Cation-induced kinetic trapping and enhanced hydrogen adsorption in a modulated anionic metal–organic framework. *Nat. Chem.*, *1*, 6, 487, 2009.
51. Zhou, X.-P., Xu, Z., Zeller, M., Hunter, A.D., Reversible uptake of $HgCl_2$ in a porous coordination polymer based on the dual functions of carboxylate and thioether. *Chem. Commun.*, 36, 5439–5441, 2009.
52. Zhang, Z., Zhang, L., Wojtas, L., Nugent, P., Eddaoudi, M., Zaworotko, M.J., Templated synthesis, postsynthetic metal exchange, and properties of a porphyrin-encapsulating metal–organic material. *J. Am. Chem. Soc.*, *134*, 2, 924–927, 2011.
53. Hermannsdörfer, J., Friedrich, M., Miyajima, N., Albuquerque, R.Q., Kümmel, S., Kempe, R., Ni/Pd@ MIL-101: Synergetische Katalyse mit kavitätenkonformen Ni/Pd-Nanopartikeln. *Angew. Chem. Int. Ed.*, *124*, 46, 11640–11644, 2012.
54. Singh, A.K. and Xu, Q., Metal–Organic Framework Supported Bimetallic Ni-Pt Nanoparticles as High-performance Catalysts for Hydrogen Generation from Hydrazine in Aqueous Solution. *ChemCatChem*, *5*, 10, 3000–3004, 2013.
55. Kleist, W., Maciejewski, M., Baiker, A., MOF-5 based mixed-linker metal–organic frameworks: Synthesis, thermal stability and catalytic application. *Thermochim. Acta*, *499*, 1–2, 71–78, 2010.
56. Cirujano, F.G., Leyva-Pérez, A., Corma, A., Llabres i Xamena, F.X., MOFs as Multifunctional Catalysts: Synthesis of Secondary Arylamines, Quinolines, Pyrroles, and Arylpyrrolidines over Bifunctional MIL-101. *ChemCatChem*, *5*, 2, 538–549, 2013.
57. Xiong, G., Yu, B., Dong, J., Shi, Y., Zhao, B., He, L.-N., Cluster-based MOFs with accelerated chemical conversion of CO_2 through C–C bond formation. *Chem. Commun.*, *53*, 44, 6013–6016, 2017.
58. Yan, Y.-T., Zhang, W.-Y., Zhang, F., Cao, F., Yang, R.-F., Wang, Y.-Y., Hou, L., Four new metal–organic frameworks based on diverse secondary building units: sensing and magnetic properties. *Chem. Commun.*, *47*, 5, 1682–1692, 2018.
59. Zhang, Z.-J., Shi, W., Niu, Z., Li, H.-H., Zhao, B., Cheng, P., Liao, D.-Z., Yan, S.-P., A new type of polyhedron-based metal–organic frameworks with interpenetrating cationic and anionic nets demonstrating ion exchange, adsorption and luminescent properties. *Chem. Commun.*, *47*, 22, 6425–6427, 2011.
60. Park, K.S., Ni, Z., Côté, A.P., Choi, J.Y., Huang, R., Uribe-Romo, F.J., Chae, H.K., O'Keeffe, M., Yaghi, O.M., Exceptional chemical and

thermal stability of zeolitic imidazolate frameworks. *Proc. Natl. Acad. Sci.*, *103*, 27, 10186–10191, 2006.
61. Wang, Z., Kravtsov, V.C., Zaworotko, M.J., Ternary Nets formed by Self-Assembly of Triangles, Squares, and Tetrahedra. *Angew. Chem. Int. Ed.*, *44*, 19, 2877–2880, 2005.
62. Zhang, M., Chen, Y.-P., Zhou, H.-C., Structural design of porous coordination networks from tetrahedral building units. *CrystEngComm*, *15*, 45, 9544–9552, 2013.
63. Deng, H., Doonan, C.J., Furukawa, H., Ferreira, R.B., Towne, J., Knobler, C.B., Wang, B., Yaghi, O.M., Multiple functional groups of varying ratios in metal-organic frameworks. *Science*, *327*, 5967, 846–850, 2010.
64. Tu, B., Pang, Q., Ning, E., Yan, W., Qi, Y., Wu, D., Li, Q., Heterogeneity within a mesoporous metal–organic framework with three distinct metal-containing building units. *J. Am. Chem. Soc.*, *137*, 42, 13456–13459, 2015.
65. Ryu, U., Jee, S., Rao, P.C., Shin, J., Ko, C., Yoon, M., Park, K.S., Choi, K.M., Recent advances in process engineering and upcoming applications of metal–organic frameworks. *Coord. Chem. Rev.*, *426*, 213544, 2020.
66. Furukawa, H., Müller, U., Yaghi, O.M., "Heterogeneity within order" in metal–organic frameworks. *Angew. Chem. Int. Ed.*, *54*, 11, 3417–3430, 2015.
67. Koh, K., Wong-Foy, A.G., Matzger, A.J., Coordination copolymerization mediated by Zn4O (CO2R) 6 metal clusters: a balancing act between statistics and geometry. *J. Am. Chem. Soc.*, *132*, 42, 15005–15010, 2010.
68. Chevreau, H., Devic, T., Salles, F., Maurin, G., Stock, N., Serre, C., Mixed-Linker Hybrid Superpolyhedra for the Production of a Series of Large-Pore Iron (III) Carboxylate Metal–Organic Frameworks. *Angew. Chem. Int. Ed.*, *52*, 19, 5056–5060, 2013.
69. Zheng, S.-T., Bu, J.T., Li, Y., Wu, T., Zuo, F., Feng, P., Bu, X., Pore space partition and charge separation in cage-within-cage indium–organic frameworks with high CO2 uptake. *J. Am. Chem. Soc.*, *132*, 48, 17062–17064, 2010.
70. Zheng, S.T., Wu, T., Irfanoglu, B., Zuo, F., Feng, P., Bu, X., Multicomponent Self-Assembly of a Nested Co24@ Co48 Metal–Organic Polyhedral Framework. *Angew. Chem. Int. Ed.*, *50*, 35, 8034–8037, 2011.
71. Schoedel, A., Cairns, A.J., Belmabkhout, Y., Wojtas, L., Mohamed, M., Zhang, Z., Proserpio, D.M., Eddaoudi, M., Zaworotko, M.J., The asc

Trinodal Platform: Two-Step Assembly of Triangular, Tetrahedral, and Trigonal-Prismatic Molecular Building Blocks. *Angew. Chem. Int. Ed.*, *52*, 10, 2902–2905, 2013.
72. Tan, Y.-X., He, Y.-P., Zhang, J., Cluster-organic framework materials as heterogeneous catalysts for high efficient addition reaction of diethylzinc to aromatic aldehydes. *Chem. Mater.*, *24*, 24, 4711–4716, 2012.
73. Li, M., Li, D., O'Keeffe, M., Yaghi, O.M., Topological analysis of metal–organic frameworks with polytopic linkers and/or multiple building units and the minimal transitivity principle. *Chem. Rev.*, *114*, 2, 1343–1370, 2013.
74. Lian, T.-T., Chen, S.-M., Wang, F., Zhang, J., Metal–organic framework architecture with polyhedron-in-polyhedron and further polyhedral assembly. *CrystEngComm*, *15*, 6, 1036–1038, 2013.
75. Chao, M.-Y., Zhang, W.-H., Lang, J.-P., Co2 and Co3 Mixed Cluster Secondary Building Unit Approach toward a Three-Dimensional Metal-Organic Framework with Permanent Porosity. *Molecules*, *23*, 4, 755, 2018.
76. Yuan, S., Qin, J.-S., Li, J., Huang, L., Feng, L., Fang, Y., Lollar, C., Pang, J., Zhang, L., Sun, D., Retrosynthesis of multi-component metal–organic frameworks. *Nat. Commun.*, *9*, 1, 808, 2018.
77. Zhai, Q.G., Bu, X., Mao, C., Zhao, X., Feng, P., Systematic and dramatic tuning on gas sorption performance in heterometallic metal–organic frameworks. *J. Am. Chem. Soc.*, *138*, 8, 2524–2527, 2016.
78. Botas, J.A., Calleja, G., Sánchez-Sánchez, M., Orcajo, M.G., Cobalt doping of the MOF-5 framework and its effect on gas-adsorption properties. *Langmuir*, *26*, 8, 5300–5303, 2010.
79. Zhou, Z.Y., Mei, L., Ma, C., Xu, F., Xiao, J., Xia, Q.B., Li, Z., A novel bimetallic MIL-101 (Cr,Mg) with high CO_2 adsorption capacity and CO_2/N_2 selectivity. *Chem. Eng. Sci.*, *147*, 109–117, 2016.
80. Botas, J.A., Calleja, G., Sánchez-Sánchez, M., Orcajo, M.G., Effect of Zn/Co ratio in MOF-74 type materials containing exposed metal sites on their hydrogen adsorption behaviour and on their band gap energy. *Int. J. Hydrogen Energy*, *36*, 17, 10834–10844, 2011.
81. Villajos, J.A., Orcajo, G., Martos, C., Botas, J. Á., Villacañas, J., Calleja, G., Co/Ni mixed-metal sited MOF-74 material as hydrogen adsorbent. *Int. J. Hydrogen Energy*, *40*, 15, 5346–5352, 2015.
82. Nevjestić, I., Depauw, H., Gast, P., Tack, P., Deduytsche, D., Leus, K., Van Landeghem, M., Goovaerts, E., Vincze, L., Detavernier, C., Van Der Voort, P., Sensing the framework state and guest molecules in

MIL-53 (Al) via the electron paramagnetic resonance spectrum of V IV dopant ions. *Phys. Chem. Chem. Phys.*, *19*, 36, 24545–24554, 2017.

83. Jiao, Y., Morelock, C.R., Burtch, N.C., Mounfield, W.P., Hungerford, J.T., Walton, K.S., Tuning the kinetic water stability and adsorption interactions of Mg-MOF-74 by partial substitution with Co or Ni. *Ind. Eng. Chem. Res.*, *54*, 12408–12414, 2015.
84. Šimėnas, M., Jee, B., Hartmann, M., Banys, J., Pöppl, A., Adsorption and desorption of HD on the metal–organic framework $Cu_{2.97}Zn_{0.03}(Btc)_2$ studied by three-pulse ESEEM spectroscopy. *J. Phys. Chem. C*, *119*, 51, 28530–28535, 2015.
85. Marti, R.M., Howe, J.D., Morelock, C.R., Conradi, M.S., Walton, K.S., Sholl, D.S., Hayes, S.E., CO2 Dynamics in Pure and Mixed-Metal MOFs with Open Metal Sites. *J. Phys. Chem. C*, *121*, 46, 25778–25787, 2017.
86. Nouar, F., Devic, T., Chevreau, H., Guillou, N., Gibson, E., Clet, G., Daturi, M., Vimont, A., Grenèche, J.M., Breeze, M.I., Walton, R.I., Tuning the breathing behaviour of MIL-53 by cation mixing. *Chem. Commun.*, *48*, 82, 10237–10239, 2012.
87. Depauw, H., Nevjestić, I., De Winne, J., Wang, G., Haustraete, K., Leus, K., Verberckmoes, A., Detavernier, C., Callens, F., De Canck, E., Vrielinck, H., Microwave induced "egg yolk" structure in Cr/V-MIL-53. *Chem. Commun.*, *53*, 60, 8478–8481, 2017.
88. Hong, K., Bak, W., Moon, D., Chun, H., Bistable and Porous Metal–Organic Frameworks with Charge-Neutral acs Net Based on Heterometallic $M_3O(CO_2)_6$ Building Blocks. *Cryst. Growth Des.*, *13*, 9, 4066–4070, 2013.
89. Lau, C.H., Babarao, R., Hill, M.R., A route to drastic increase of CO_2 uptake in Zr metal organic framework UiO-66. *Chem. Commun.*, *49*, 35, 3634–3636, 2013.
90. Song, X., Oh, M., Lah, M.S., Hybrid bimetallic metal–organic frameworks: modulation of the framework stability and ultralarge CO_2 uptake capacity. *Inorg. Chem.*, *52*, 19, 10869–10876, 2013.
91. Wei, Z., Lu, W., Jiang, H.L., Zhou, H.C., A route to metal–organic frameworks through framework templating. *Inorg. Chem.*, *52*, 3, 1164–1166, 2013.
92. Zheng, S.T., Wu, T., Chou, C., Fuhr, A., Feng, P., Bu, X., Development of composite inorganic building blocks for MOFs. *J. Am. Chem. Soc.*, *134*, 10, 4517–4520, 2012.
93. Dinca, M. and Long, J.R., High-enthalpy hydrogen adsorption in cation-exchanged variants of the microporous metal– organic framework

$Mn_3[(Mn_4Cl)_3(BTT)_8(CH_3OH)_{10}]_2$. *J. Am. Chem. Soc.*, *129*, 36, 11172–11176, 2007.

94. Kaur, G., Rai, R.K., Tyagi, D., Yao, X., Li, P.Z., Yang, X.C., Zhao, Y., Xu, Q., Singh, S.K., Room-temperature synthesis of bimetallic Co–Zn based zeolitic imidazolate frameworks in water for enhanced CO_2 and H_2 uptakes. *J. Mater. Chem. A*, *4*, 39, 14932–14938, 2016.
95. Liao, J.H., Chen, W.T., Tsai, C.S., Wang, C.C., Characterization, adsorption properties, metal ion-exchange and crystal-to-crystal transformation of $Cd_3[(Cd_4Cl)_3(BTT)_8(H_2O)_{12}]_2$ framework, where BTT^{3-}= 1, 3, 5-benzene tristetrazolate. *CrystEngComm*, *15*, 17, 3377–3384, 2013.
96. Mendt, M., Jee, B., Himsl, D., Moschkowitz, L., Ahnfeldt, T., Stock, N., Hartmann, M., Pöppl, A., A continuous-wave electron paramagnetic resonance study of carbon dioxide adsorption on the metal–organic frame-work MIL-53. *Appl. Magn. Reson.*, *45*, 3, 269–285, 2014.
97. Hong, K., Bak, W., Chun, H., Unique coordination-based heterometallic approach for the stoichiometric inclusion of high-valent metal ions in a porous metal–organic framework. *Inorg. Chem.*, *52*, 10, 5645–5647, 2013.
98. Smith, S.J., Ladewig, B.P., Hill, A.J., Lau, C.H., Hill, M.R., Post-synthetic Ti exchanged UiO-66 metal-organic frameworks that deliver exceptional gas permeability in mixed matrix membranes. *Sci. Rep.*, *5*, 1, 1–6, 2015.
99. Ferrando-Soria, J., Serra-Crespo, P., de Lange, M., Gascon, J., Kapteijn, F., Julve, M., Cano, J., Lloret, F., Pasán, J., Ruiz-Pérez, C., Journaux, Y., Selective gas and vapor sorption and magnetic sensing by an isoreticular mixed-metal–organic framework. *J. Am. Chem. Soc.*, *134*, 37, 15301–15304, 2012.
100. Cao, Y., Zhao, Y., Song, F., Zhong, Q., Alkali metal cation doping of metal-organic framework for enhancing carbon dioxide adsorption capacity. *J. Energy Chem.*, *23*, 4, 468–474, 2014.
101. Timofeeva, M.N., Panchenko, V.N., Khan, N.A., Hasan, Z., Prosvirin, I.P., Tsybulya, S.V., Jhung, S.H., Isostructural metal-carboxylates MIL-100(M) and MIL-53(M) (M: V, Al, Fe and Cr) as catalysts for condensation of glycerol with acetone. *Appl. Catal. A: Gen.*, *529*, 167–174, 2017.
102. Fu, Y., Xu, L., Shen, H., Yang, H., Zhang, F., Zhu, W., Fan, M., Tunable catalytic properties of multi-metal–organic frameworks for aerobic styrene oxidation. *Chem. Eng. J.*, *299*, 135–141, 2016.
103. Mitchell, L., Williamson, P., Ehrlichová, B., Anderson, A.E., Seymour, V.R., Ashbrook, S.E., Acerbi, N., Daniels, L.M., Walton, R.I., Clarke,

M.L., Wright, P.A., Mixed-Metal MIL-100 (Sc,M)(M= Al, Cr, Fe) for Lewis Acid Catalysis and Tandem C=C Bond Formation and Alcohol Oxidation. *Chem.-Eur. J.*, *20*, 51, 17185–17197, 2014.

104. Sun, D., Sun, F., Deng, X., Li, Z., Mixed-metal strategy on metal–organic frameworks (MOFs) for functionalities expansion: Co substitution induces aerobic oxidation of cyclohexene over inactive Ni-MOF-74. *Inorg. Chem.*, *54*, 17, 8639–8643, 2015.
105. Saha, D., Hazra, D.K., Maity, T., Koner, S., Heterometallic Metal–Organic Frameworks That Catalyze Two Different Reactions Sequentially. *Inorg. Chem.*, *55*, 12, 5729–5731, 2016.
106. Nouar, F., Breeze, M.I., Campo, B.C., Vimont, A., Clet, G., Daturi, M., Devic, T., Walton, R.I., Serre, C., Tuning the properties of the UiO-66 metal organic framework by Ce substitution. *Chem. Commun.*, *51*, 77, 14458–14461, 2015.
107. Wen, M., Kuwahara, Y., Mori, K., Zhang, D., Li, H., Yamashita, H., Synthesis of Ce ions doped metal–organic framework for promoting catalytic H_2 production from ammonia borane under visible light irradiation. *J. Mater. Chem. A*, *3*, 27, 14134–14141, 2015.
108. Schejn, A., Aboulaich, A., Balan, L., Falk, V., Lalevée, J., Medjahdi, G., Aranda, L., Mozet, K., Schneider, R., Cu^{2+}-doped zeolitic imidazolate frameworks (ZIF-8): efficient and stable catalysts for cycloadditions and condensation reactions. *Catal. Sci. Technol.*, *5*, 3, 1829–1839, 2015.
109. Abbasi, A., Soleimani, M., Najafi, M., Geranmayeh, S., New interpenetrated mixed (Co/Ni) metal–organic framework for dye removal under mild conditions. *Inorg. Chim. Acta*, *439*, 18–23, 2016.
110. Krap, C.P., Newby, R., Dhakshinamoorthy, A., García, H., Cebula, I., Easun, T.L., Savage, M., Eyley, J.E., Gao, S., Blake, A.J., Lewis, W., Enhancement of CO_2 adsorption and catalytic properties by Fe-doping of $[Ga_2(OH)_2(L)](H_4L$=biphenyl-3, 3′, 5, 5′-tetracarboxylic acid), MFM-300 (Ga_2). *Inorg. Chem.*, *55*, 3, 1076–1088, 2016.
111. Pal, T.K., De, D., Neogi, S., Pachfule, P., Senthilkumar, S., Xu, Q., Bharadwaj, P.K., Significant Gas Adsorption and Catalytic Performance by a Robust CuII–MOF Derived through Single-Crystal to Single-Crystal Transmetalation of a Thermally Less-Stable ZnII–MOF. *Chem.-Eur. J.*, *21*, 52, 19064–19070, 2015.
112. Zhang, Z., Zhang, L., Wojtas, L., Nugent, P., Eddaoudi, M., Zaworotko, M.J., Templated synthesis, postsynthetic metal exchange, and properties of a porphyrin-encapsulating metal–organic material. *J. Am. Chem. Soc.*, *134*, 2, 924–927, 2012.

113. Aguirre-Díaz, L.M., Gándara, F., Iglesias, M., Snejko, N., Gutiérrez-Puebla, E., Monge, M.Á., Tunable catalytic activity of solid solution metal-organic frameworks in one-pot multicomponent reactions. *J. Am. Chem. Soc.*, *137*, 19, 6132–6135, 2015.
114. Zhao, C., Dai, X., Yao, T., Chen, W., Wang, X., Wang, J., Yang, J., Wei, S., Wu, Y., Li, Y., Ionic exchange of metal–organic frameworks to access single nickel sites for efficient electroreduction of CO_2. *J. Am. Chem. Soc.*, *139*, 24, 8078–8081, 2017.
115. Sun, D., Liu, W., Qiu, M., Zhang, Y., Li, Z., Introduction of a mediator for enhancing photocatalytic performance via post-synthetic metal exchange in metal–organic frameworks (MOFs). *Chem. Commun.*, *51*, 11, 2056–2059, 2015.
116. Lee, Y., Kim, S., Kang, J.K., Cohen, S.M., Photocatalytic CO_2 reduction by a mixed metal (Zr/Ti), mixed ligand metal–organic framework under visible light irradiation. *Chem. Commun.*, *51*, 26, 5735–5738, 2015.
117. Amador, R.N., Carboni, M., Meyer, D., Sorption and photodegradation under visible light irradiation of an organic pollutant by a heterogeneous UiO-67–Ru–Ti MOF obtained by post-synthetic exchange. *RSC Adv.*, *7*, 1, 195–200, 2017.
118. Tu, J., Zeng, X., Xu, F., Wu, X., Tian, Y., Hou, X., Long, Z., Microwave-induced fast incorporation of titanium into UiO-66 metal–organic frameworks for enhanced photocatalytic properties. *Chem. Commun.*, *53*, 23, 3361–3364, 2017.
119. Rasero-Almansa, A.M., Iglesias, M., Sánchez, F., Synthesis of bimetallic Zr(Ti)-naphthalendicarboxylate MOFs and their properties as Lewis acid catalysis. *RSC Adv.*, *6*, 108, 106790–106797, 2016.
120. Vu, T.A., Le, G.H., Dao, C.D., Dang, L.Q., Nguyen, K.T., Dang, P.T., Tran, H.T., Duong, Q.T., Nguyen, T.V., Lee, G.D., Isomorphous substitution of Cr by Fe in MIL-101 framework and its application as a novel heterogeneous photo-Fenton catalyst for reactive dye degradation. *RSC Adv.*, *4*, 78, 41185–41194, 2014.
121. Wang, X.S., Chrzanowski, M., Wojtas, L., Chen, Y.S., Ma, S., Corrigendum: Formation of a Metalloporphyrin-Based Nanoreactor by Postsynthetic Metal–Ion Exchange of a Polyhedral-Cage Containing a Metal–Metalloporphyrin Framework. *Chem.-Eur. J.*, *19*, 37, 12187–12187, 2013.

6
Complexity Based on Chiral Framework—Part 1

Abstract
This chapter presents the various synthetic approaches in the formation of chiral MOFs through direct method and/or inherently due to using chiral component (enantiopure ligand as starting agent) or helical backbone production. Complexity can be derived from the many interesting chiral features in MOFs. These kinds of chiral MOFs can be considered as an important subclass of chiral complex materials with various applications due to their properties. So, chiral structural design and chiralization strategy are very significant factors in this field.

Keywords: Asymmetric synthesis, direct chiralization, inherent chirality, racemic, enantioselective applications

6.1 Inherent Chirality

Chirality is a fundamental characteristic and common phenomenon in chemistry, biology, physics, and science of materials. By comparison to different asymmetric compounds such as chiral chemical materials (zeolites, MOFs, metal oxides, and activated carbon as porous carbon), chiral MOFs have been attracted much attention. Since, chiral MOFs have variety and changeable appropriate structural properties and they can be considered as a major part of chiral materials. Their applications have been also studied in academy even industry, particularly processes of biological and pharmaceutical [1–4]. For the first time, a homochiral MOF catalyst was reported to be used in the asymmetric catalytic reactions in 2000 by Kim *et al.*

Ali Morsali and Kayhaneh Berijani. Metal-Organic Frameworks with Heterogeneous Structures, (105–126) © 2021 Scrivener Publishing LLC

Its structure consisted of a chiral framework that was prepared by tartaric acid derivative and zinc nitrate hexahydrate. Day to day, the considerable progresses are observed in the chiral MOFs field that are impetus to generate new chiral MOFs with various designs. In recent years, a number of homochiral metal-organic frameworks have been synthesized and applied in wide range of asymmetric applications [5–11]. Surely, the application of chiral MOFs is due to chemical composition, structural flexibility in internal surface area, porosity, and skeleton, and the generation of chiral environment [12]. The asymmetric application is very important to produce the enantiomerical enriched compounds particularly drugs. Since, drugs are chiral and they have two enantiomers. It is important that both of the enantiomers are not helpful as drug, for example, thalidomide R enantiomer has drug capability but S enantiomer causes the birth defect [13]. The other applications have been discussed in details in the next sections. A review has been also reported about the enantioselective catalysis with chiral MOFs [14]. As well, metal-camphorate frameworks as chiral MOFs were discussed to form review by Bu *et al.* [15]. In asymmetric catalysis field with CMOFs, we published new works with simple design, recently [16–18]. Chirality in porous solids such as MOFs is very important from the point of two aspects: first, the components of MOF that can be chiral, and second, the induction of chirality to framework that can be done with the arrangement of their components (chiral MOF built from achiral components). Actually, there are several methods to produce chiral MOFs such as chiral ligands or chiral templates, chiral physical environment, and spontaneous resolution without any chiral auxiliary during crystallization process [19–22]. But specifically, chiralization of framework in MOF (homochiral) can be classified to the three general approaches: a) achiral metal precursors with enantiopure linkers, b) achiral node with combination of enantiopure and achiral linkers, and c) chiral metal precursors (achiral node + chiral auxiliary) and achiral linkers. Each of these approaches has advantages to chiralize MOFs and the simplest characterization is CD spectra that reveal chirality behavior. In the following, we classify this section in details: inherent chirality, direct synthesis that can be performed through chiral linker and node, chiral template synthesis, and chiral post-synthetic modification. Truly, the study and investigation in the production of chiral MOFs are hot and important topic for having unique properties. Unfortunately, this issue has been dealt less because of the existence of complexities and synthetic challenging.

In addition to the progress in the syntheses of chiral MOFs by using chiral components, the chiral spontaneous resolution with achiral components based on crystallization, crystal packing, is a hard and complex strategy [23, 24]. In chiral MOFs that are produced from achiral precursors without any chiral species, chiral space group, can generate spontaneous resolution in structure. The crystals of different MOFs can be formed in a space group that is chiral and it causes the spatial deformation of structural units [25]. SBUs units in 3D frameworks can be chiral due to the difference of the connections between organic linkers and metal centers that lead to chiral SBUs (Figures 6.1a, b) [26]. N2 adsorption isotherm of complex 1 shows a type I sorption behavior. The apparent surface area was also calculated by using the Langmuir method to be 479 $m^2 g^{-1}$ which confirmed the permanent porosity of the considered structure. The methane adsorption isotherm shows type-I sorption behavior with an uptake of 2.5 wt % at 80 bar and 295 K. The obtained results were suitable. It should be noted that Complex1crystallizes in point group Cs, belonging to one of the 10 polar point groups (C_1, C_s, C_2, C_2V, C_4, C_4V, C_3, C_3V, C_6, and C_6V). Ferroelectricity of this compound in single crystal form was also investigated. Totally, this 3D microporous framework showed chiral tetrahedral SBUs along with asymmetric properties as ferroelectric and NLO properties.

Authors claimed that their work may exploit easy entry for incorporation multifunctionalities into MOFs for the preconceived requires in designs of material.

Inherent chirality can be observed in MOFs crystalline with helical morphology. Achiral building units can have C2 space group that leads to distorted helix. The helix structure in the framework of MOFs can be an effective source in the generation of chirality. 3D metal-organic framework can be synthesized with helical chains by using an achiral ligand [27]. Clearly, the configuration around inorganic nodes (metal nodes) can be chiral without any chiral precursor that can generate homochiral frameworks through spontaneous resolution [28–31]. However, this kind of chiralization is a rare

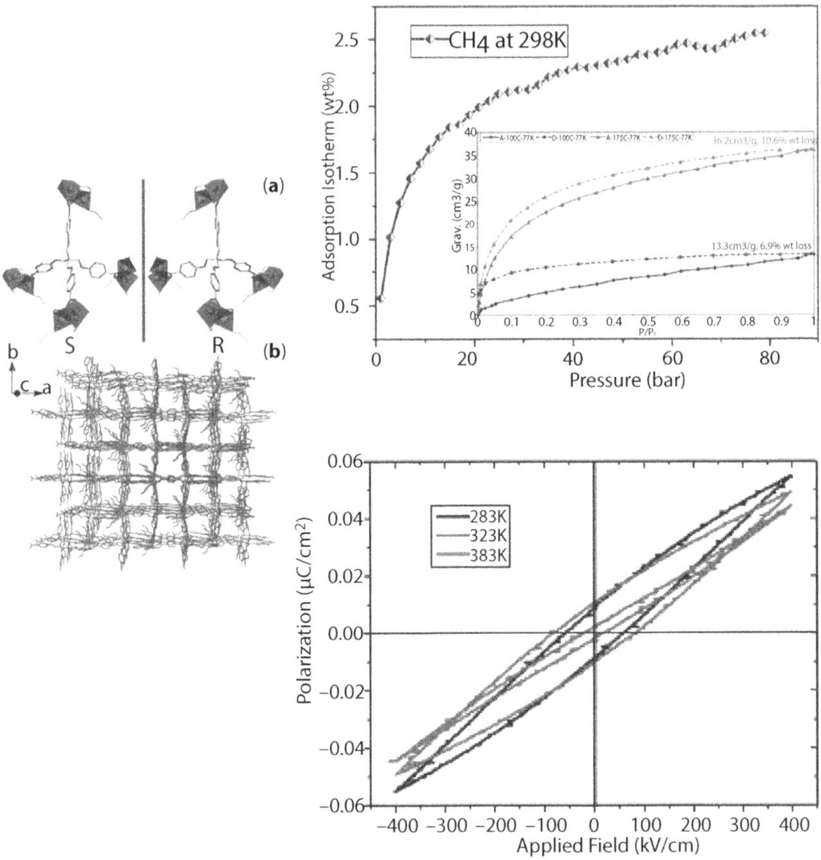

Figure 6.1 (a) Two types of chiral tetrahedral SBUs. (b) A perspective view of microporous framework: zinc, azury; carbon, gray; oxygen, red. Water molecules of lattice are eliminate for clarity (left). Adsorption isotherm of methane at 298 K, and H_2 adsorption desorption isotherms (77 K, inset) activated at 100°C (blue) and 175°C (red) (right and top). Polarization (P) versus applied field (E) hysteresis loops from a single crystal sample (right and bottom) [26].

event compared to the other ways and it is still a big challenge [32]. Some examples were reported about homochiral MOFs by crystallization from achiral precursors but the control of enantioriched single crystals is very hard. Unfortunately, the spontaneous resolution method in preparing chiral MOFs leads to racemic kinds (heterochiral) [33]. The racemic MOFs can be converted to homochiral MOFs with less changing, for example, in a pillared layer metal-organic framework (PL-MOF), one type of racemic chiral pillar was converted to its homochiral which can be along with a pillar pattern changing [34]. The intrinsic chirality is also observed in the framework of zeolites but they are often racemic so the construction of enantioenriched zeolites is a very difficult experiment [35–37]. Due to nature of chiral compounds, complexity in their structure is clearly observed. For example, using interweaving multi-helices is another synthetic strategy in the creation of chiral MOFs. There are various reports in this field, for instance, Hou *et al.* reported a chiral MOF by using diisophthalate ligand that has pyridyl site. There are interweaving parallel four-fold left-handed and two-fold right-handed helical chains in the considered MOF structure that synthesize heterochiral six-fold helices and consequently 1D cylindrical chiral channel along the c-axis is prepared. It can be said that although the most of the time L^{4-} is in planner mode and, in turn, formed achiral framework, but basis of chirality in this context can be attributed to L^{4-} molecule bending with the intention of helix formation [38]. Totally, the use of chiral component to obtain of chiral MOFs is logical, very effective, and useful method rather than the other methods that researchers prefer them.

6.2 Direct Chirality

In the synthesis of MOFs, the type of used components is main factor from solvent to the framework components. Major contribution of information in literature from the past until now about CMOFs obtained from chiral components: chiral organic linker and chiral nodes, so we start discussion with chiral linkers. Using the chiral linkers especially enantiopure in the framework of MOFs is a key selection in the creation of chiral crystalline MOF. The most important chiral ligands that are used in the synthesis of chiral materials are such as BINAP (2,2′-bis(diphenylphosphino)-1,1′-binaphthyl), BINOL (1,1′-Bi-2-naphthol), BOX (bis(oxazoline)), and SALEN. Indeed, the connection of homochiral ligands to metallic units generates a chiral porous material. Even, the numbers of homochiral chains can participate in the creation of CMOFs. This direct way is one case of synthetic chiralization with chiral component in the framework that can lead to the structural complexity [39–42]. Chiral ligands can be central or axial ligands based on the type of chirality

[43]. In the central chiral ligand method, a ligand chiral as amino-acids or carboxylic acids can be used in both of ionic and neutral frameworks directly as $Cu_2(D\text{-cam})_2(4,4'\text{-bpy})$ [34], $Ag_4(\text{triazole})_6$ [44], $[Zn_3(m_3\text{-}O)(L_2)6](H_3O)_2{\cdot}12(H_2O)$, POST-1 ($HL_2$: tartaric acid based ligand) [15], and $[Zn_2(\beta\text{-alanyl-L-histidine})_2]{\cdot}6H_2O$ (analogue of ZIFs family) [45]. In addition to the generated chiral environment in pores or cavities in MOFs, the chiral components can have asymmetric capabilities. They can be considered an ideal platform to synthesize heterogenized asymmetric catalysts owing to their ability to generate uniform active catalytic sites with identical secondary environments around them throughout the solid. Also, they can have unprecedentedly high loadings of catalyst centers, more accessible active sites, and prolonged catalyst life times since the elimination of multi molecular catalyst-deactivation pathways. According to the size of chiral channel and the existence of functional groups, the active metal sites can be immobilized on linker as (BINOLate) $Ti(O^iPr)_2$ (Schematic representation, Figure 6.2a) [46]. 3D MOF had

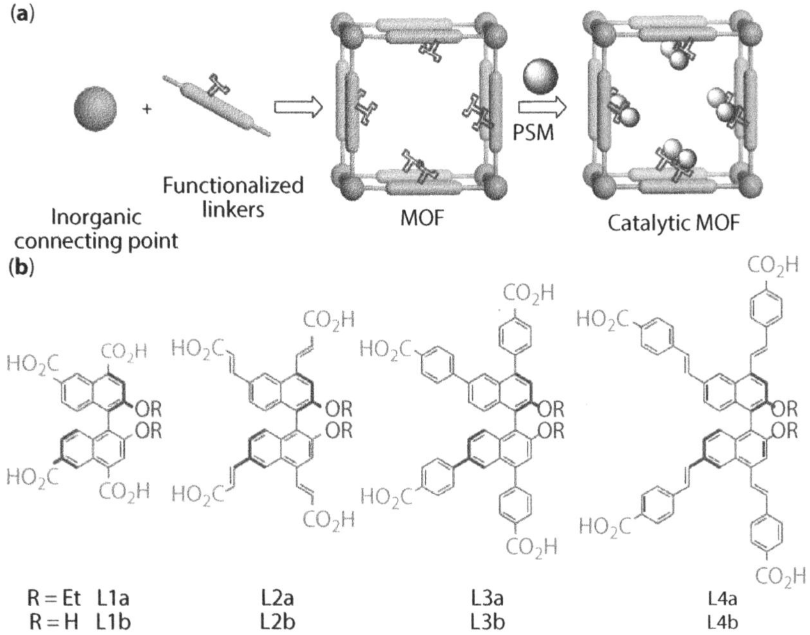

Figure 6.2 (a) Schematic representation of a chiral MOF catalyst by PSM. (b) The chemical structures of the used chiral ligands based on tetracarboxylic acids [46].

suitable channels with the hydroxyl groups that their metalation was done by using $Ti(O^iPr)_4$. This MOF plus other seven MOFs in this series were designed and synthesize with the framework formula $[LCu_2(solvent)_2]$ (L: chiral tetracarboxylate ligand derived from 1,1′-bi-2-naphthol). A series of tetracarboxylic acids with orthogonal chiral diethoxy (L1a to L4a) or dihydroxy (L1b to L4b) functional groups in (R)-enantiomeric form were synthesized for this project (Figure 6.2b).

They have the same architectures with specific crystal structures (such as CMOF-1a) but channels with different sizes (Figure 6.3). Certainly, the presence of various pore sizes can produce complexity in MOF structure, easily. After structural investigations, catalytic activity of isoreticular CMOFs was also studied. They have the same non-interpenetrating framework structures and due to different open channel sizes, they can provide an ideal platform for

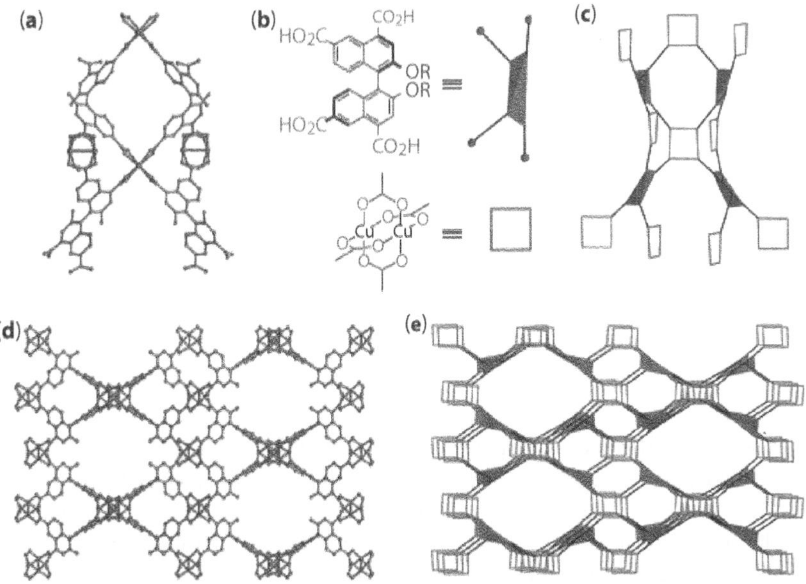

Figure 6.3 Crystal structure of chiral MOF-1a. (a) Paddle-wheels clusters based on Cu and their connectivity with chiral L1a ligands. (b) L1a: blue distorted tetrahedro; Cu-paddle-wheel cluster: red square. (c) Connectivity of L1a organic ligands and Cu-paddle-wheel node. (d) Stick model. (e) Representation of a simplified connectivity (top) [46]. *(Continued)*

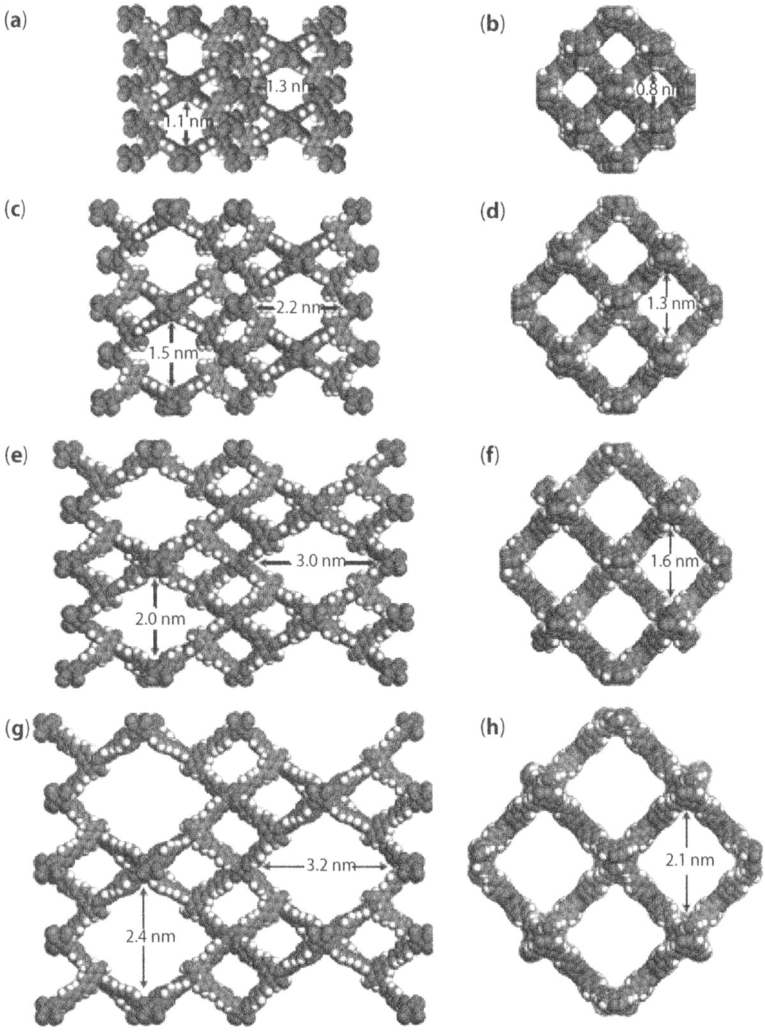

Figure 6.3 (Continued) Crystal structure of chiral MOF-1a. (a) Paddle-wheels clusters based on Cu and their connectivity with chiral L1a ligands. (b) L1a: blue distorted tetrahedro; Cu-paddle-wheel cluster: red square. (c) Connectivity of L1a organic ligands and Cu-paddle-wheel node. (d) Stick model. (e) Representation of a simplified connectivity (top). (a–h) Space-filling models of CMOF-1b to -4b with pores size (bottom) [46].

asymmetric catalysis like diethylzinc and ethyl(phenylalkynyl)zinc addition to several aldehydes for production of chiral secondary alcohols (Tables 6.1 and 6.2).

Table 6.1 shows that CMOF-3b was used to catalyze with conversions up to.99%, excellent selectivity to main products, secondary

Table 6.1 Diethylzinc additions to aromatic aldehydes.

Entry	CMOF-	Aromatic group	Selectivity (%)	Conversion (%)	ee (%)
1	**3b**	1-Naph	92	>99	91
2	**3b**	4-Cl-Ph	84	>99	80
3	**3b**	4-Br-Ph	70	96	80
4	**3b**	4-Me-Ph	74	>99	78
5	**3b**	Ph	68	>99	82
6	**L3b-Me$_4$**	1-Naph	70	94	94
7	**L3b-Me$_4$**	4-Cl-Ph	>99	>99	88
8	**L3b-Me$_4$**	4-Br-Ph	99	>99	87
9	**L3b-Me$_4$**	4-Me-Ph	>99	81	87
10	**L3b-Me$_4$**	Ph	>99	>99	86
11	**4b**	Ph	82	>99	84
12	**2b**	Ph	64	98	70
13	**1b**	Ph	91	98	<3
14	**3a**	Ph	10	>99	0
15	**4b**	1-Naph	70	60	91
16	**2b**	1-Naph	87	>99	86
17	**1b**	1-Naph	95	81	9
18	**3b**	Ph	51	>99	66
19	**3b**	Ph	93	>99	72
20	**3b**	Ph	97	97	67
21	**3b**	Ph	98	93	68

(Continued)

Table 6.1 Diethylzinc additions to aromatic aldehydes. (*Continued*)

Entry	CMOF-	Aromatic group	Selectivity (%)	Conversion (%)	ee (%)
22	**3b**	Ph	98	87	57
23	**3b**	Ph	99	99	64

The additions were catalyzed by the CMOF/Ti(O^{i}Pr)$_4$ combinations. All the reactions were carried out with vigorous stirring with a magnetic stir bar except those for entries 18–23 in which the mixtures were agitated only gently to avoid pulverizing the solid catalysts and to facilitate the recovery of CMOFs. This experimental deviation gave rise to different levels of enantiomeric excesses since diffusion of the substrates in entries 18–23 is slower so the contribution from the non-enantioselective background reaction is higher. 1-Naph, 1-naphthaldehyde.

Table 6.2 Alkynylzinc additions to aromatic aldehydes (similar conditions with Table 6.1).

Entry	CMOF-	Aromatic group	Conversion (%)	ee (%)
1	**3b**	1-Naph	>99	71
2	**3b**	4-Cl-Ph	>99	59
3	**3b**	4-Br-Ph	>99	69
4	**3b**	4-Me-Ph	>99	31
5	**3b**	Ph	>99	76
6	**1b**	Ph	>99	0
7	**2b**	Ph	>99	0
8	**4b**	Ph	>99	77
9	**4b**	1-Naph	>99	49
10	**3a**	Ph	>99	0

alcohol, up to 99% and enantiomeric excess for 1-naphthaldehyde up to 91%. The other results in tables show that these chiral MOF as asymmetric heterogeneous catalysts is very good. Also, the sizes of the open channels in the CMOF series were investigated in substrate and product diffusion, as shown in Table 6.1 (entries 5, 11–13). For example, 1-phenyl-1-propanol ee was dependent on channel sizes of the considered chiral MOF. CMOF-1b/Ti(OiPr)$_4$ gave 3% ee that can be related to small channel size that cannot accommodate both reagents. To show effective synthesis of CMOFs and their catalytic performance, CMOF-3b and Ti(OiPr)$_4$ combination was employed for ethyl(phenylalkynyl)zinc and various aldehydes addition. Complete conversions and 76% ee to chiral alcohols were seen (Table 6.2). All of these results show that these MOFs with heterogeneous structures and complexity in their body due to different open channel sizes can show high performance in asymmetric catalysis. So, there is the competition between design and complexity degree. It can be predicted that the whole dihydroxyl groups have not been metalated and the topology of the MOF affects enantioselectivity of a specific product. A new view of complexity in 3D MOF in addition to the chiral ligand is the changing in the dimension of the chiral helical chains that can happen [47]. Wu *et al.* reported five chiral metal-organic coordination polymers synthesis with high thermal stability through hydro(solvo)thermal method. The used components were: divalent metal ions as node (Co^{2+}, Zn^{2+}, and Cd^{2+}), D-camphoric acid as enantiopure organic linker, and achiral neutral N,N′-bis(pyrid-4-yl)piperazine ligand (bpypip). Scheme 6.1 shows

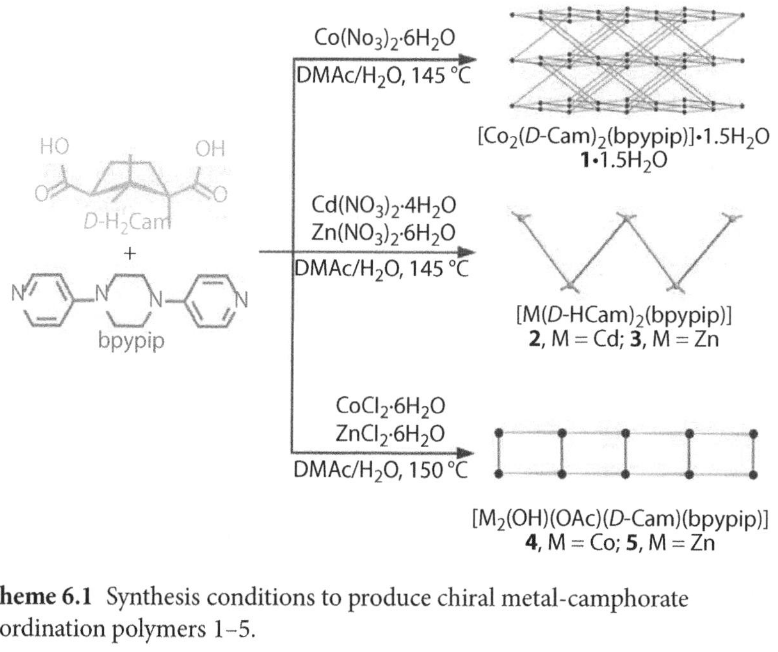

Scheme 6.1 Synthesis conditions to produce chiral metal-camphorate coordination polymers 1–5.

hydro(solvo)thermal synthesis conditions for the production of chiral metal-camphorate coordination polymers 1–5.

In Scheme 6.2, coordination modes of D-Cam and D-HCam organic ligands in these polymers 1–5 have been shown. It is well known that carboxylate group can domain diversified coordination modes. Scheme 6.2 shows two carboxylate groups of D-Cam ligand in two different coordination modes that one adopts a bidentate *syn,syn*-bridging mode while another is in the form of a bidentate chelating-bridging mode. Figure 6.4 from a to c shows Λ- and Δ-configuration, anti relationship that is observed between two α-methyl groups of D-HCam with respect to the central Cd^{2+} and 1D zigzag coordination chain structure. These synthesized chiral materials have been successfully synthesized through the use of enantioselective and dual-ligand synthesis methods simultaneously.

The introduction of achiral N-tethering polytopic ligands into the homochiral metal-camphorate architectures undoubtedly shows the generation of new polymeric structures with different network topology that show various level of structural complexity.

Another chiral ligand in framework can be axial chiral ligand. In this case, chiral ligands are synthesized with atropisomeric biaryl linkers as binaphthyl derivatives that have the substituent groups and are used in the construction of CMOFs [48–57]. The structure of final chiral ligand can coordinate to different metals and this metalloligand in framework can have the different abilities as asymmetric catalysis [58]. However, in comparison to central ligands to construct CMOFs, axial ligands have been less used [59]. For example, Lin *et al.* reported isoreticular chiral MOFs that had cubic network

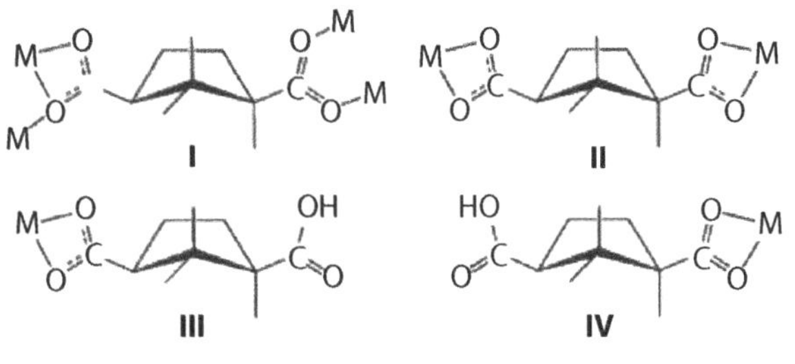

Scheme 6.2 Coordination modes of D-Cam and D-HCam in polymeric structures 1–5.

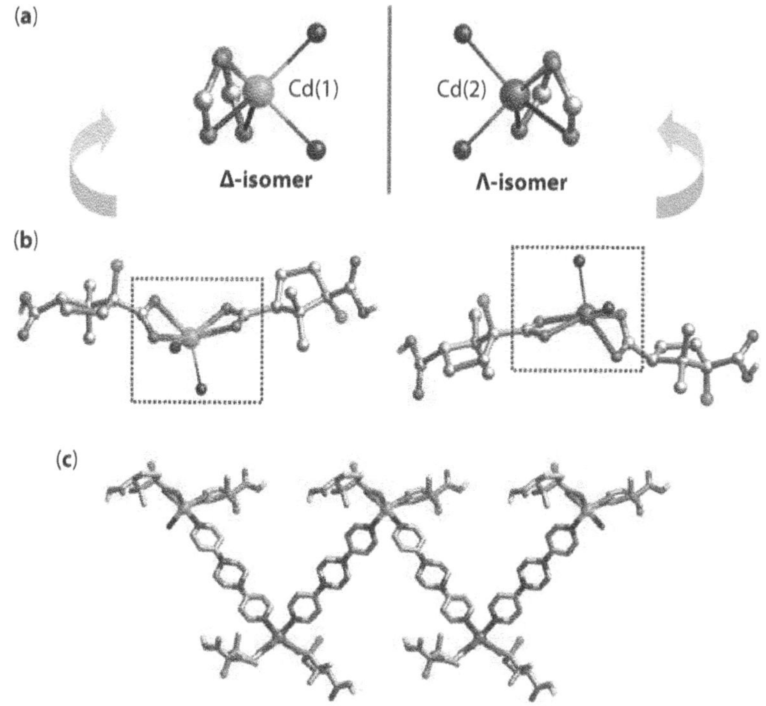

Figure 6.4 (a and b) Views of connectivity around metallic centers in structure 2, highlighting D,L-enantiomeric Cd2+ ions. (c) The one-dimensional zigzag coordination chain structure in 2. Cyan: Cd(1); pink: Cd(2) atom; red: oxygen atom; blue: nitrogen atom; gray: carbon atom; yellow: hydrogen atom. α-methyl groups of D-HCam organic ligands: green-colored balls (similar to Zn-analogous 3) [47].

topology. They were synthesized by using $[Zn_4(\mu_4\text{-}O)(O_2CR)_6]$ secondary building units and dicarboxylate struts from chiral Mn-Salen family under solvothermal method (Figure 6.5).

CMOFs 1-5 were investigated by several techniques as TGA, XRD, and PXRD. Interestingly, some of these frameworks with different pore sizes (Figure 6.6) explicitly include two-fold or non-interpenetrated structures. The appearance of this phenomenon depends on solvents steric sizes that were employed to synthesis MOF crystals. Three-fold interpenetrated framework was also observed owing to Mn-Salen-strut that has extreme length (Figure 6.7). In these cases, the existence of the various open channel and pore sizes in frameworks can be considered complexity factor. These structures

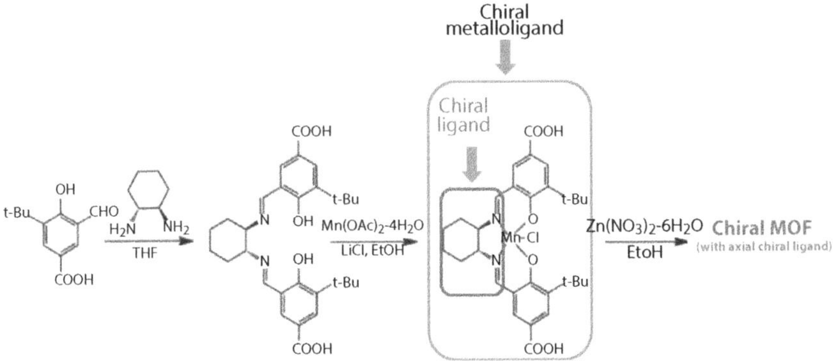

Figure 6.5 Total synthesis of CMOF with chiral axial ligand [59].

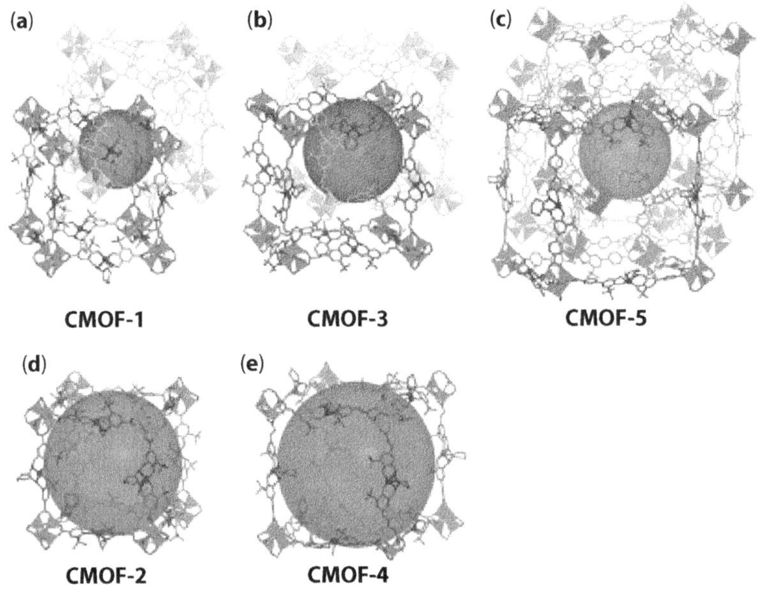

Figure 6.6 Different cavity sizes in CMOFs [58].

due to three different chiral Mn-Salen struts in their structure could show catalytic activity in the enantioselective epoxidation unfunctionalized olefins with high enantiomeric excess. Olefin epoxidation rate strongly depends on channel sizes and activity of catalytic centers. These results suggest that diffusion of substrate and oxidant in catalytic reactions depend on channels of catalyst. Incorporation of chiral complexes in the struts that their investigation is difficult causes that the organic linker to be chiral [60, 61]. As mentioned,

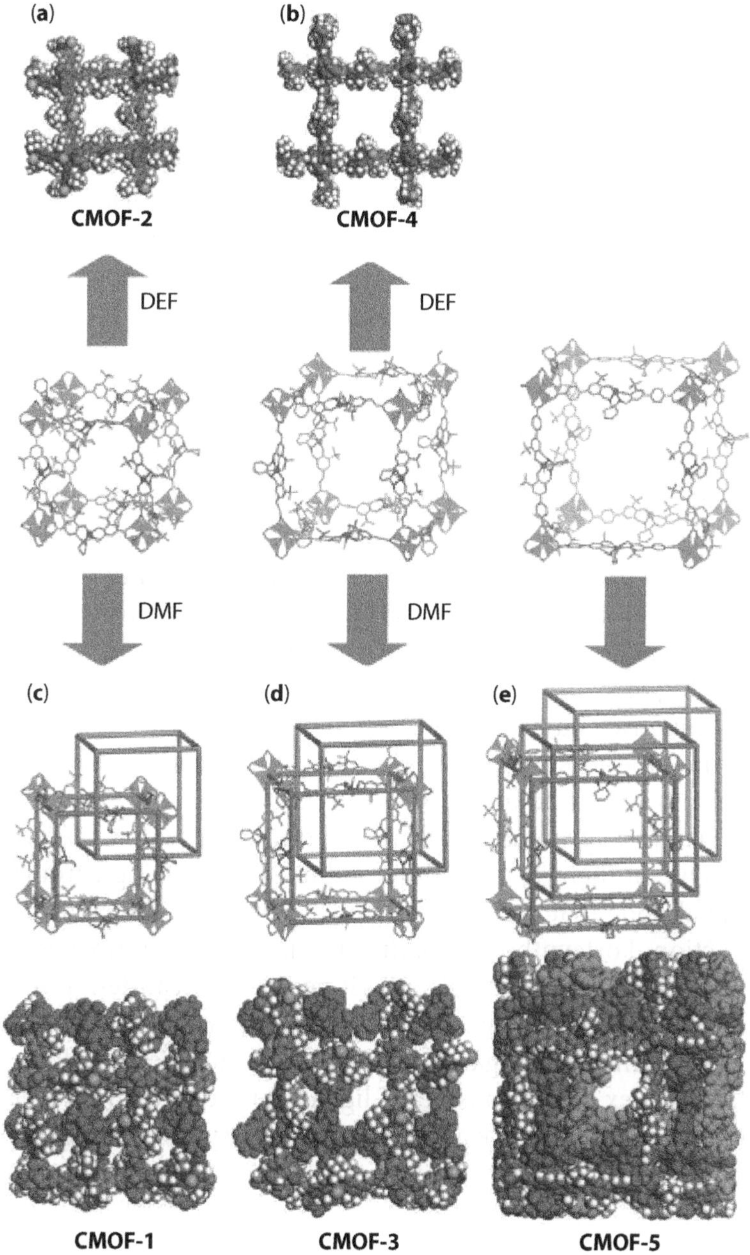

Figure 6.7 Stick/polyhedra models of CMOF-1-5 with the presence of interpenetration along with connectivity of ligands and SBUs [66].

the presence of mixed ligands in MOFs can generate complexity in framework. The mixed linker method in the chiral systems is observed, too.

When mixed chiral/achiral organic linker is employed in the construction of CMOFs, certainly, the effect of chiral linker and chiral induction will decrease but the change in the dimension and type of linkers can assist to produce the structural complexity [47, 62]. The observations show that chiral MOFs develop toward chiral SURMOFs that have more significant applications [63]. Chiral thin film can be formed from two moieties that in one of them, one kind of chiral organic linkers has been employed. This change induces heterogeneity to system that affects the structural complexity of final material. Until now, the remarkable CMOFs have been synthesized but their preparation methods are still challenging. In addition to the important factor in the creation CMOFs, chiral organic linkers, the asymmetric metal nodes can also induce chirality into topology of MOFs. The symmetry at metal node can be decreased due to the attachment of linker and/or incorporation of a species so with decreasing of symmetry, complexity can increase. Metal cluster can be asymmetric and used in the design of CMOFs.

In the other case, the used clusters can be synthesized by chiral ligand and employed in the preparation of chiral MOFs [64]. Some strategies were presented to chiralize frameworks that can be efficient. One of them is postchiral modification by using chiral agent on metal node that causes MOF node to be chiral due to acid-base interaction. Postsynthetic modification is one of the three essential methods to fabricate CMOFs and it generates ideal design for this kind of synthesis. The chiral node can be chiralized due to the incorporation of the chiral ligands *via* postchiral modification. Even, the metal complex of the considered ligand can be directly attached to metal node of framework and they can have asymmetric capabilities [65].

6.3 Conclusion

There are different strategies in MOFs chiralization that inherent and direct methods can be introduced. In the first method, twisting and chiral helical chains are usually observed. Indeed, helicity in structure can produce chirality and induce chiroptical activity. However, this kind of chirality is not enough in material in the effective chirality induction. It should be noted that chiral space groups can also generate chirality in materials. In the second method, using enantiopure species in the generation of chiral MOFs is an important factor that they can be chiral linker or chiral cluster. Of course, the number of chiral MOFs based on chiral linkers is greater than based on chiral cluster owing to convenience of their synthesis. Amonoacids, Salen, BINOL, and 1,1′-biphenol are famous chiral linkers that are used in the synthesis of CMOFs. Another point about chiral MOFs is that they can have various applications with asymmetric direction that some of them in this chapter have been explained. Chirality due to its incentive motivational properties is an important agent to motivate scientists that investigate various chiral MOFs by using different synthetic methods and their uses. So, due to chiral MOFs importance, the other cases have been also introduced in the next chapter.

References

1. Eliel, E.L., Wilen, S.H., Mander, L.N., *Stereochemistry of Organic Compounds*, vol. 13, pp. 991–1118, Wiley & Sons, Inc., New York, Cap, 1994.
2. Wang, Y., Xu, J., Wang, Y., Chen, H., Emerging chirality in nanoscience. *Chem. Soc. Rev.*, *42*, 7, 2930–2962, 2013.
3. Farina, V., Reeves, J.T., Senanayake, C.H., Song, J.J., Asymmetric synthesis of active pharmaceutical ingredients. *Chem. Rev.*, *106*, 7, 2734–2793, 2006.
4. Stinson, S.C., Chiral pharmaceuticals. *Chem. Eng. News*, *79*, 79–97, 2001.
5. Maurin, G., Serre, C., Cooper, A., Férey, G., The new age of MOFs and of their porous-related solids. *Chem. Soc. Rev.*, *46*, 11, 3104–3107, 2017.
6. Zhou, H.-C. and Kitagawa, S., Metal-organic frameworks (MOFs). *Chem. Soc. Rev.*, *43*, 5415–5418, 2014.
7. Zhou, H.-C., Long, J.R., Yaghi, O.M., Introduction to metal-organic frameworks. *Chem. Rev.*, *112*, 673–674, 2012.

8. Bradshaw, D., Claridge, J., Cussen, E., Prior, T., Rosseinsky, M.J., Design, chirality, and flexibility in nanoporous molecule-based materials. *Acc. Chem. Res.*, *38*, 4, 273–282, 2005.
9. Kesanli, B. and Lin, W., Chiral porous coordination networks: rational design and applications in enantioselective processes. *Coord. Chem. Rev.*, *246*, 1–2, 305–326, 2003.
10. Dybtsev, D.N., Nuzhdin, A.L., Chun, H., Bryliakov, K.P., Talsi, E.P., Fedin, V.P., Kim, K., A homochiral metal–organic material with permanent porosity, enantioselective sorption properties, and catalytic activity. *Angew. Chem. Int. Ed.*, *45*, 6, 916–920, 2006.
11. Zhang, W. and Xiong, R.-G., Ferroelectric metal–organic frameworks. *Chem. Rev.*, *112*, 2, 1163–1195, 2011.
12. Bhattacharjee, S., Khan, M.I., Li, X., Zhu, Q.-L., Wu, X.-T., Recent progress in asymmetric catalysis and chromatographic separation by chiral metal–organic frameworks. *Catalysts*, *8*, 3, 120, 2018.
13. Jacques, V., Czarnik, A.W., Judge, T.M., Van der Ploeg, L.H., DeWitt, S.H., Differentiation of antiinflammatory and antitumorigenic properties of stabilized enantiomers of thalidomide analogs. *Proc. Natl. Acad. Sci.*, *112*, 12, E1471–E1479, 2015.
14. Horcajada, P., Salles, F., Wuttke, S., Devic, T., Heurtaux, D., Maurin, G., Vimont, A., Daturi, M., David, O., Magnier, E., How linker's modification controls swelling properties of highly flexible iron (III) dicarboxylates MIL-88. *J. Am. Chem. Soc.*, *133*, 44, 17839–17847, 2011.
15. Seo, J.S., Whang, D., Lee, H., Im Jun, S., Oh, J., Jeon, Y.J., Kim, K., A homochiral metal–organic porous material for enantioselective separation and catalysis. *Nature*, *404*, 6781, 982, 2000.
16. Berijani, K. and Morsali, A., Dual activity of durable chiral hydroxyl-rich MOF for asymmetric catalytic reactions. *J. Catal.*, *378*, 28–35, 2019.
17. Berijani, K., Morsali, A., Hupp, J.T., An Effective Strategy for Creating Asymmetric MOFs for Chirality Induction: A Chiral Zr-based MOF for Enantioselective Epoxidation. *Catal. Sci. Technol.*, *9*, 3388–3397, 2019.
18. Razavi, S.A.A., Berijani, K., Morsali, A., Hybrid nanomaterials for asymmetric purposes: green enantioselective C–C bond formation by chiralization and multi-functionalization approaches. *Catal. Sci. Technol.*, 10, *24*, 8240–8253, 2020.
19. Wu, C.-D., Hu, A., Zhang, L., Lin, W., A homochiral porous metal–organic framework for highly enantioselective heterogeneous asymmetric catalysis. *J. Am. Chem. Soc.*, *127*, 25, 8940–8941, 2005.

20. Katsuki, I., Motoda, Y., Sunatsuki, Y., Matsumoto, N., Nakashima, T., Kojima, M., Spontaneous resolution induced by self-organization of chiral self-complementary cobalt (III) complexes with achiral tripod-type ligands containing three imidazole groups. *J. Am. Chem. Soc.*, *124*, 4, 629–640, 2002.
21. Li, J.R., Tao, Y., Yu, Q., Bu, X.H., Sakamoto, H., Kitagawa, S., Selective gas adsorption and unique structural topology of a highly stable guest-free zeolite-type MOF material with N-rich chiral open channels. *Chem.-Eur. J.*, *14*, 9, 2771–2776, 2008.
22. Siemeling, U., Scheppelmann, I., Neumann, B., Stammler, A., Stammler, H.-G., Frelek, J., Spontaneous chiral resolution of a coordination polymer with distorted helical structure consisting of achiral building blocks. *Chem. Commun.*, 17, 2236–2237, 2003.
23. Eliel, E.L. and Kofron, J.T., The Resolution of p-Ethylphenylmethylcarbinol. Infrared Spectra of Enantiomorphs and Racemates. *J. Am. Chem. Soc.*, *75*, 18, 4585–4587, 1953.
24. Addadi, L. and Lahav, M., Chemistry, A., Towards the planning and execution of an "absolute" asymmetric synthesis of chiral dimers and polymers with quantitative enantiomeric yield. *Pure Appl. Chem.*, *51*, 6, 1269–1284, 1979.
25. Moulton, B. and Zaworotko, M.J., From molecules to crystal engineering: supramolecular isomerism and polymorphism in network solids. *Chem. Rev.*, *101*, 6, 1629–1658, 2001.
26. Guo, Z., Cao, R., Wang, X., Li, H., Yuan, W., Wang, G., Wu, H., Li, J., A multifunctional 3D ferroelectric and NLO-active porous metal–organic framework. *J. Am. Chem. Soc.*, *131*, 20, 6894–6895, 2009.
27. Tian, G., Zhu, G., Yang, X., Fang, Q., Xue, M., Sun, J., Wei, Y., Qiu, S., A chiral layered Co (II) coordination polymer with helical chains from achiral materials. *Chem. Commun.*, 11, 1396–1398, 2005.
28. Pérez-García, L. and Amabilino, D.B., Spontaneous resolution under supramolecular control. *Chem. Soc. Rev.*, *31*, 6, 342–356, 2002.
29. Katsuki, I., Motoda, Y., Sunatsuki, Y., Matsumoto, N., Nakashima, T., Kojima, M., Spontaneous resolution induced by self-organization of chiral self-complementary cobalt (III) complexes with achiral tripod-type ligands containing three imidazole groups. *J. Am. Chem. Soc.*, *124*, 4, 629–640, 2002.
30. Zhang, J., Chen, S., Nieto, R.A., Wu, T., Feng, P., Bu, X.A., Tale of Three Carboxylates: Cooperative Asymmetric Crystallization of a Three-Dimensional Microporous Framework from Achiral Precursors. *Angew. Chem. Int. Ed.*, *49*, 7, 1267–1270, 2010.

31. Jeong, K.U., Yang, D.K., Graham, M.J., Tu, Y., Kuo, S.W., Knapp, B.S., Harris, F.W., Cheng, S.Z., Construction of chiral propeller architectures from achiral molecules. *Adv. Mater.*, *18*, 24, 3229–3232, 2006.
32. Jacques, J., Collet, A., Wilen, S.H., *Enantiomers, racemates and resolution*, Krieger Publishing Company, Florida, 1994.
33. Zhang, S.-Y., Li, D., Guo, D., Zhang, H., Shi, W., Cheng, P., Wojtas, L., Zaworotko, M.J., Synthesis of a chiral crystal form of MOF-5, CMOF-5, by chiral induction. *J. Am. Chem. Soc.*, *137*, 49, 15406–15409, 2015.
34. Zhang, J., Yao, Y.-G., Bu, X., Comparative study of homochiral and racemic chiral metal-organic frameworks built from camphoric acid. *Chem. Mater.*, *19*, 21, 5083–5089, 2007.
35. Sun, J., Bonneau, C., Cantín, Á., Corma, A., Díaz-Cabañas, M.J., Moliner, M., Zhang, D., Li, M., Zou, X., The ITQ-37 mesoporous chiral zeolite. *Nature*, *458*, 7242, 1154, 2009.
36. Rojas, A. and Camblor, M.A., A pure silica chiral polymorph with helical pores. *Angew. Chem. Int. Ed.*, *51*, 16, 3854–3856, 2012.
37. Tong, M., Zhang, D., Fan, W., Xu, J., Zhu, L., Guo, W., Yan, W., Yu, J., Qiu, S., Wang, J., Synthesis of chiral polymorph A-enriched zeolite Beta with an extremely concentrated fluoride route. *Sci. Rep.*, *5*, 11521, 2015.
38. Liu, B., Zhou, H.F., Hou, L., Zhu, Z., Wang, Y.Y., A chiral metal–organic framework with polar channels: unique interweaving six-fold helices and high CO_2/CH_4 separation. *Inorg. Chem. Front.*, *3*, 10, 1326–1331, 2016.
39. Du, D.-Y., Qin, J.-S., Li, S.-L., Su, Z.-M., Lan, Y.-Q., Recent advances in porous polyoxometalate-based metal–organic framework materials. *Chem. Soc. Rev.*, *43*, 13, 4615–4632, 2014.
40. Shimizu, G.K., Vaidhyanathan, R., Taylor, J.M., Phosphonate and sulfonate metal organic frameworks. *Chem. Soc. Rev.*, *38*, 5, 1430–1449, 2009.
41. Farha, O.K. and Hupp, J.T., Rational design, synthesis, purification, and activation of metal- organic framework materials. *Acc. Chem. Res.*, *43*, 8, 1166–1175, 2010.
42. Wu, T., Zhang, J., Zhou, C., Wang, L., Bu, X., Feng, P., Zeolite RHO-type net with the lightest elements. *J. Am. Chem. Soc.*, *131*, 17, 6111–6113, 2009.
43. Han, Z., Shi, W., Cheng, P., Synthetic strategies for chiral metal-organic frameworks. *Chin. Chem. Lett.*, *29*, 6, 819–822, 2018.
44. Iremonger, S.S., Southon, P.D., Kepert, C.J., A nanoporous chiral metal–organic framework material that exhibits reversible guest adsorption. *Dalton Trans.*, 44, 6103–6105, 2008.

45. Katsoulidis, A.P., Park, K.S., Antypov, D., Martí-Gastaldo, C., Miller, G.J., Warren, J.E., Robertson, C.M., Blanc, F., Darling, G.R., Berry, N.G., Guest-Adaptable and Water-Stable Peptide-Based Porous Materials by Imidazolate Side Chain Control. *Angew. Chem. Int. Ed.*, *53*, 1, 193–198, 2014.
46. Ma, L., Falkowski, J.M., Abney, C., Lin, W., A series of isoreticular chiral metal–organic frameworks as a tunable platform for asymmetric catalysis. *Nat. Chem.*, *2*, 10, 838–846, 2010.
47. Wu, J.-Y. and Huang, S.-M., Homochiral transition-metal camphorate coordination architectures containing "piperazine–pyridine" ligands. *CrystEngComm*, *13*, 6, 2062–2070, 2011.
48. Jeong, K.S., Go, Y.B., Shin, S.M., Lee, S.J., Kim, J., Yaghi, O.M., Jeong, N., Asymmetric catalytic reactions by NbO-type chiral metal–organic frameworks. *Chem. Sci.*, *2*, 5, 877–882, 2011.
49. Lin, W., Asymmetric catalysis with chiral porous metal–organic frameworks. *Top. Catal.*, *53*, 13–14, 869–875, 2010.
50. Bringmann, G., Price Mortimer, A.J., Keller, P.A., Gresser, M.J., Garner, J., Breuning, M., Atroposelective synthesis of axially chiral biaryl compounds. *Angew. Chem. Int. Ed.*, *44*, 34, 5384–5427, 2005.
51. Ma, L., Wu, C.D., Wanderley, M.M., Lin, W., Single-Crystal to Single-Crystal Cross-Linking of an Interpenetrating Chiral Metal–Organic Framework and Implications in Asymmetric Catalysis. *Angew. Chem. Int. Ed.*, *49*, 44, 8244–8248, 2010.
52. Zheng, M., Liu, Y., Wang, C., Liu, S., Lin, W., Cavity-induced enantioselectivity reversal in a chiral metal–organic framework Brønsted acid catalyst. *Chem. Sci.*, *3*, 8, 2623–2627, 2012.
53. Ma, L., Mihalcik, D.J., Lin, W., Highly porous and robust 4, 8-connected metal–organic frameworks for hydrogen storage. *J. Am. Chem. Soc.*, *131*, 13, 4610–4612, 2009.
54. Wanderley, M.M., Wang, C., Wu, C.-D., Lin, W., A chiral porous metal–organic framework for highly sensitive and enantioselective fluorescence sensing of amino alcohols. *J. Am. Chem. Soc.*, *134*, 22, 9050–9053, 2012.
55. Yuan, G., Zhu, C., Liu, Y., Xuan, W., Cui, Y., Anion-driven conformational polymorphism in homochiral helical coordination polymers. *J. Am. Chem. Soc.*, *131*, 30, 10452–10460, 2009.
56. Hu, A., Ngo, H.L., Lin, W., Chiral porous hybrid solids for practical heterogeneous asymmetric hydrogenation of aromatic ketones. *J. Am. Chem. Soc.*, *125*, 38, 11490–11491, 2003.

57. Ngo, H.L. and Lin, W., Chiral crown ether pillared lamellar lanthanide phosphonates. *J. Am. Chem. Soc.*, *124*, 48, 14298–14299, 2002.
58. Song, F., Wang, C., Falkowski, J.M., Ma, L., Lin, W., Isoreticular chiral metal-organic frameworks for asymmetric alkene epoxidation: tuning catalytic activity by controlling framework catenation and varying open channel sizes. *J. Am. Chem. Soc.*, *132*, 43, 15390–15398, 2010.
59. Tan, X., Zhan, J., Zhang, J., Jiang, L., Pan, M., Su, C.-Y., Axially chiral metal-organic frameworks produced from spontaneous resolution with an achiral pyridyl dicarboxylate ligand. *CrystEngComm*, *14*, 1, 63–66, 2012.
60. Shultz, A.M., Farha, O.K., Adhikari, D., Sarjeant, A.A., Hupp, J.T., Nguyen, S.T., Selective surface and near-surface modification of a noncatenated, catalytically active metal-organic framework material based on Mn(salen) struts. *Inorg. Chem.*, *50*, 8, 3174–3176, 2011.
61. Song, F., Wang, C., Lin, W., A chiral metal-organic framework for sequential asymmetric catalysis. *Chem. Commun.*, *47*, 29, 8256–8258, 2011.
62. Haldar, R. and Maji, T.K., Metal-organic frameworks (MOFs) based on mixed linker systems: structural diversities towards functional materials. *CrystEngComm*, *15*, 45, 9276–9295, 2013.
63. Liu, B., Shekhah, O., Arslan, H.K., Liu, J., Wöll, C., Fischer, R.A., Enantiopure metal-organic framework thin films: oriented SURMOF growth and enantioselective adsorption. *Angew. Chem. Int. Ed.*, *51*, 3, 807–810, 2012.
64. Rood, J.A., Noll, B.C., Henderson, K.W., Homochiral frameworks derived from magnesium, zinc and copper salts of L-tartaric acid. *J. Solid State Chem.*, *183*, 1, 270–276, 2010.
65. Horcajada, P., Salles, F., Wuttke, S., Devic, T., Heurtaux, D., Maurin, G., Vimont, A., Daturi, M., David, O., Magsnier, E., How linker's modification controls swelling properties of highly flexible iron (III) dicarboxylates MIL-88. *J. Am. Chem. Soc.*, *133*, 44, 17839–17847, 2011.
66. Falkowski, J.M., Liu, S., Wang, C. and Lin, W., Chiral metal-organic frameworks with tunable open channels as single-site asymmetric cyclopropanation catalysts. *Chem. Commun.*, *48*, 52, 6508–6510, 2012.

7

Complexity Based on Chiral Framework—Part 2

Abstract

The other strategies for production of chiral MOFs are chiral-template and post-synthesis. They can happen to different models, for example, chiral or achiral template can generate chiral centers in MOFs. Also, post-chiral modification is an attractive method to synthesize CMOFs that are achiral frameworks. So, selection of the chiral molecule in these syntheses is key parameter.

The synthesized materials by these methods can be complex along with diverse structural properties. So, in this chapter, the generated complexities *via* these synthetic methods have been investigated and their effects on applications.

Keywords: Post-chiral modification, indirect chiralization, chiral-template synthesis, enantiopure ligands, achiral frameworks, enantioselective applications

7.1 Chiral-Template Synthesis

However, there are complexities in asymmetric syntheses particularly in CMOFs, but still the chiral nature is attractive and valuable. So, these materials are currently attracting considerable attention in view of the asymmetric applications. To synthesize this kind of material, network structure and its chemical nature can be manipulated. However, there is difficulty in preparing chiral MOFs without enantiomer intergrowth within single crystallites. So, using chiral ligand in structures is very effective method. In addition to the direct synthesis

Ali Morsali and Kayhaneh Berijani. Metal-Organic Frameworks with Heterogeneous Structures, (127–148) © 2021 Scrivener Publishing LLC

of chiral MOFs, their synthesis can be performed by using the chiral auxiliary ligand or template that has been known as indirect chiralization toward the production of chiral environment. The chiral template can be an enantiopure ligand that induces chirality into achiral MOF framework with specific enantiomeric case, and finally, CMOF will be constructed [1]. The control of helices handedness and direction of the growth of crystals toward chiral network are due to chiral template. Some of the chiral templates are such as L- or D-N-tert-butoxycarbonyl-2-(imidazole)-1-pyrrolidone, enantiopure 1,2-propanediol, (−)-cinchonidine, (+)-cinchonine, and L-proline [1–4]. For instance, the structure determination of the 1,2-propanediol (1,2-pd) phase by Rosseinsky *et al.* represents the first demonstration that chiral molecules can specifically template helix handedness in a chiral porous framework solid. This group reported that to search for a suitable template to grow homochiral bulk samples, methyl-substituted chiral glycol 1,2-propanediol could be used as chiral solvent. This change in the synthesis process leads to chiral phase due to a pronounced distortion of the net. The origin of the chiral templating effect has been shown in Figure 7.1. According to observations, diol methyl group points away from framework and into helical pore structure. Chiral center inversion of 1,2-pd would bring methyl group into excessive

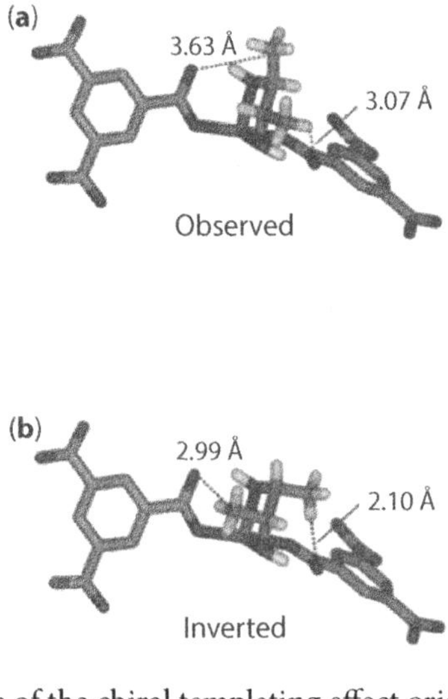

Figure 7.1 An example of the chiral templating effect origin [1].

proximity with btc unit of framework, which is unfavorable. For observed enantiomer the closest contacts with carboxylate nonbonding oxygen are longer that they have been demonstrated in Figure 7.1 than opposite enantiomer. But, the presence of strong templating interaction is due to below distances of framework-template.

The chiral templates can be counter ion: cationic and anionic chiral templates [5, 6]. For example, Duan and co-workers reported CMOFs to enantiomorph form by using Keggin-type anion, asymmetric organocatalytic group (PYI) along with nickel-source. One of the important issues in these chiral heterogeneous materials is effect of self-assembly method. Ni-PYIs channels were enlarged through a guest exchange reaction to remove the cationic chiral templates.

These materials could show high activity in the asymmetric dihydroxylation of aryl olefins (Figure 7.2). For the first case, it was supposed that using POM can provide redox activity due to the presence of surface oxygen. This agent along with catalytic sites of chiral PYIs and hydrophilic/hydrophobic features these MOFs are significant parameters in the considered asymmetric catalysis.

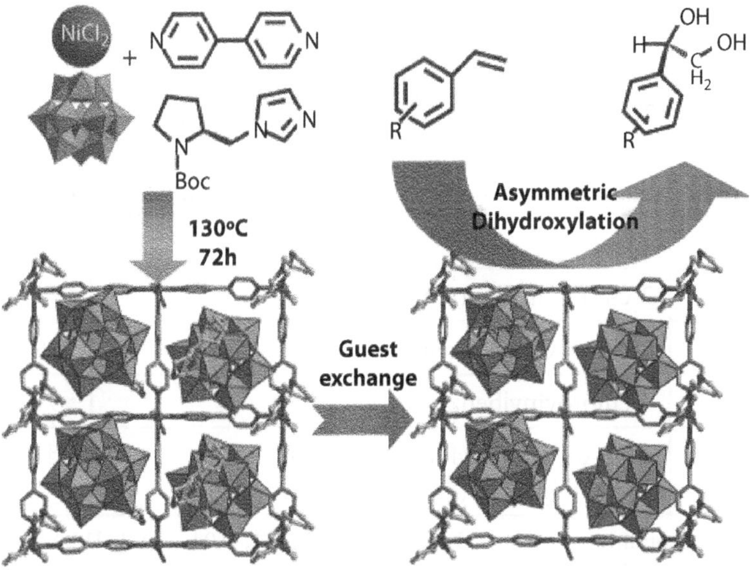

Figure 7.2 Synthesis method of Ni-PYI1, guest exchange, and the considered asymmetric catalysis [46].

After structural characterizations such as powder x-ray diffraction, single-crystal, and PLATON analyses, aryl olefins asymmetric dihydroxylation was investigated and their yields and enantiomeric excess have been demonstrated in Table 7.1. Excellent enantioselectivity was a good result for Ni-PYI1 and also its effect in catalytic reaction was studied *via* filtration test. It showed that conversion after filtration has been increased only 5%.

It seems that ionic chiral template can be a kind of metal complexes that is chiral and has the chirality induction possibility to framework [7–9]. It is expected that the content of chirality in environment is related to the amount of chiral counter ions in ionic MOF structure. Importantly, enantioselective sensing of chiral molecules can be performed *via* ion-exchange by using chiral ionic MOFs [6]. Also, ionic liquids can be chiral and used as chiral template in the creation of CMOFs. The adding of ionic liquid or the simple chiral template as proline into MOF can also generate distorted skeleton in MOFs that leads to the complexity of framework. That is interesting that achiral templates can also provide chiral positions in MOFs

Table 7.1 Asymmetric dihydroxylation of aryl olefins.[a]

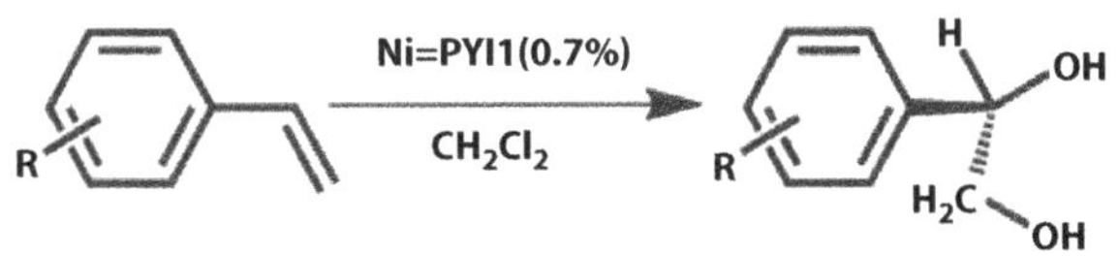

Entry	Substrate	Conversion (%)[b]	ee (%)[c]
1	styrene (1)	75	>95
2	2-chlorovinylbenzene (2)	76	67
3	3-chlorovinylbenzene (3)	79	>95
4	4-chlorovinylbenzene (4)	75	>95
5	5 3,5-di-tert-butyl-4'-vinylbiphenyl (5)	<10	nd

[a]Reaction conditions: olefin, 55 mmol; *ex*-Ni-PYI1, 0.04 mmol; H2O2 (15%), 15 ml; CH_2Cl_2, 5 ml; 40°C; 60 h.
[b]The conversions were determined by 'H NMR spectroscopy of crude products.
[c]The ee value was determined bv chiral HPLC on a Chiralcel OD-H column.

although the choosing of achiral templates is an important aspect [10]. The indirect synthesis of CMOFs can be carried out in the presence of another present chiral component in environment as solvent and catalyst that can induce chirality to surrounding [11–14].

7.2 Post-Synthesis

The post-chiral modification can happen about organic linkers of framework, simply. With this method (and the other methods), the functionalization of MOFs can be done by specific chiral species for determined purposes [15–22]. In fact, this process leads to new CMOFs. Cohen and co-workers reported many investigations about PSM in MOFs. One of their works was chiralization of achiral IRMOF-3 by using chiral alkyl anhydrides such as (S)-(+)-2-methylbutyric anhydride or cyclic anhydride (S)-(-)-2 acetoxy-succinic anhydride. Degree of reaction conversion was determined by using ^{1}H NMR analysis through digestion method. Interestingly, nature of modifying agent is very effective on the conversion to chiral IRMOF-3 [23].

Chiral modification of linker is a best method for production of the particular features in structures of MOFs like complexity. The simplest chiral PSM can happen on functional frameworks such as MIL-101-NH_2. Although, its chiral counterpart is used in chiral gas chromatography but our goal from the discussion of this example is showing chiral MOFs that are produces with post-chiral modification. Recently, a MOF from MILs' family was selected with amine functional group to immobilize of chiral compounds through PSM strategy. The reason of selection of MIL-101(Al)-NH_2 as achiral parent MOF is its useful features [24]. The versions of new CMIL generated conditions to chiral separation of the different racemates species based on the present interactions (Figure 7.3) [25]. However, there is still limited availability about CMOFs, their synthesis, recognition and applications. Yang, Yan, and co-workers reported five chiral MOFs with same parent framework and different chiral species that were grafted. Their diversity in the used chiral functional groups can affect chiral selectivity and separation for chiral chromatography. Five chiral molecules with different chiral groups or chiral centers

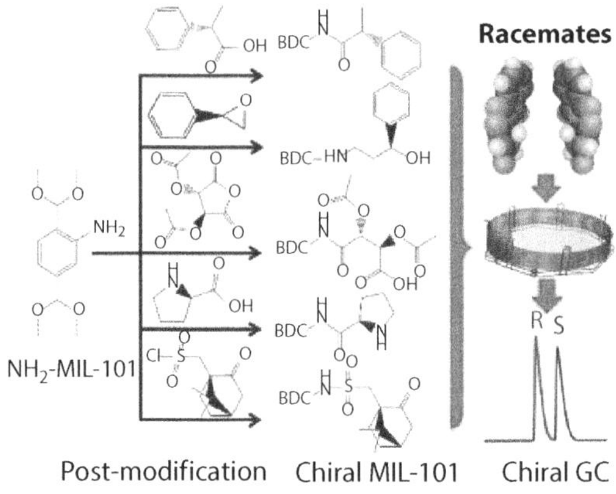

Figure 7.3 Examples of the post-chiral modification of achiral MOF-NH_2 as capillary columns for chiral GC [25].

were used ((S)-2-Phenylpropionic acid, (R)-1,2-epoxyethylbenzene, (+)-diacetyl-L-tartaric anhydride, L-proline, and (1S)-(+)-10-camphorsulfonyl chloride).

It is important that the chiral selectivity of CMOFs is different. This finding can be related to post-modify chiral groups' nature and their effective interactions with racemate (Figure 7.4). In addition to the simple functional groups on organic linker and easy interaction with chiral agents, the other functions can exist on achiral frameworks as alkynes that can be functionalized by chiral species. The obtained chiral reticular MOFs can have the ability in the asymmetric catalysis (Figure 7.5) [26–29].

For more understanding of this issue, two enantiomeric Zn-MOFs can be introduced that have two kinds of proline enantiomers and they were synthesized *via in situ* click reactions by modification of two adducts with opposite chirality within the same achiral MOFs, respectively. The used compounds to prepare two enantiomeric Zn-MOF1 and Zn-MOF2 are: dimethyl-5-(prop-2-ynyl.oxy)isophthalic acid (H_2DPYI), 4,4-dipyridine, zinc salt, and methanol by solvothermal process. This method was high-yield preparation to synthesis Zn-DPYI. This framework was crystallized in an achiral space group (Figure 7.6, left).

Each zinc was coordinated by two oxygen in one bidentate carboxyl group, two monodentate carboxyl groups from two different DPYI ligands, and two nitrogen donors from two different

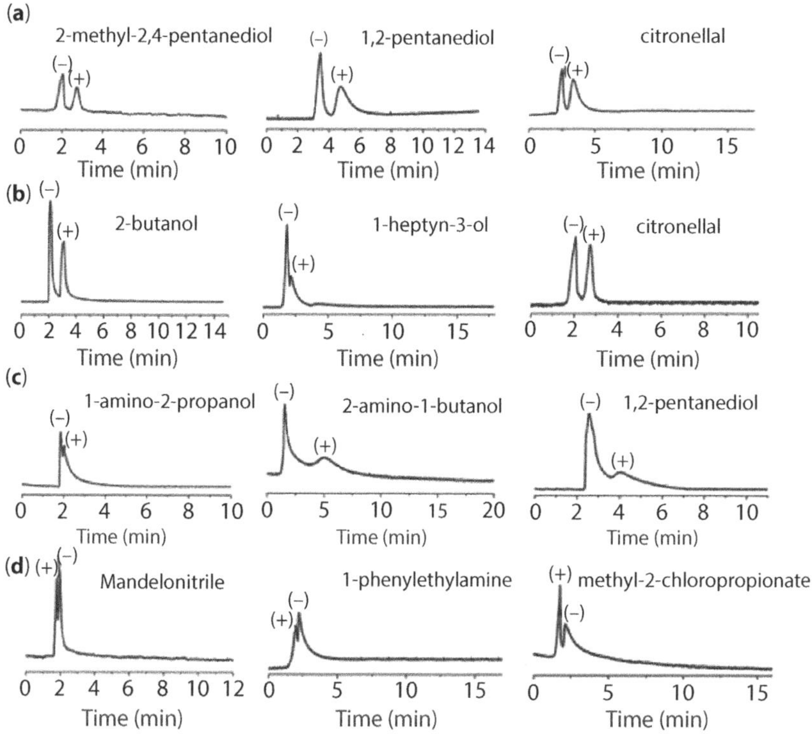

Figure 7.4 (a–d) GC chromatograms based on chiral MOF-based columns to separate racemates under different conditions. (a) MIL-101-S-2-Ppa coated column A: 2-methyl-2,4-pentanediol (160°C, 2 ml min^{-1} N_2); 1,2-pentanediol (160°C, 2 ml min^{-1} N_2); citronellal (165°C, 2 ml min^{-1} N_2). (b) MIL-101-R-Epo coated column B: 2-butanol (150°C, 2 ml min^{-1} N_2); 1-heptyn-3-ol (200°C, 2 ml min^{-1} N_2); citronellal (180°C, 2 ml min^{-1} N_2). (c) MIL-101-(+)-Ac-L-Ta coated column C: 1-amino-2-propanol (210°C, 2 ml min^{-1} N_2); 2-amino-1-butanol (210°C, 2 ml min^{-1} N_2); 1,2-pentanediol (210°C, 3 ml min^{-1} N_2). (d) MIL-101-L-Pro coated column D: Mandelonitrile (200°C, 1.8 ml min^{-1} N_2); 1-phenylethylamine (180°C, 2 ml min^{-1} N_2); methyl-2-chloropropionate (180°C, 2 ml min^{-1} N_2) [25].

4,4-dipyridine linkers. Two zinc atoms were connected by two bridged COO^- moieties into a dimeric moiety. Finally, produced sheets are two-dimensional that were stacked parallel to *bc* plane *via* weak interactions forming a non-interpenetrating channel-like framework with cross-section along the *a*-axis. Crystal structure of Zn-DPYI demonstrating clickable alkyne moieties positions that

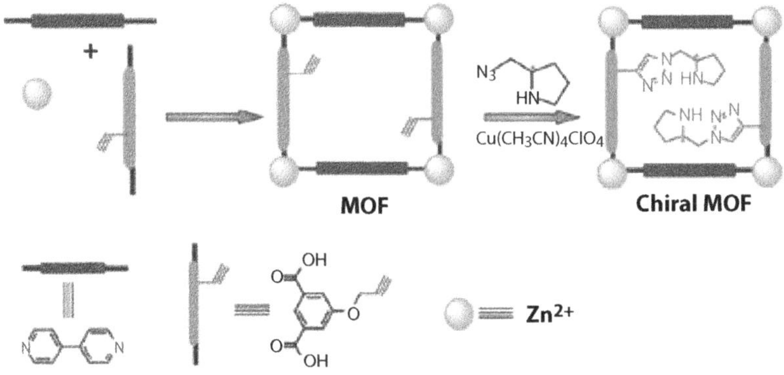

Figure 7.5 The example of modified MOF with chiral species through post-chiral modification [26].

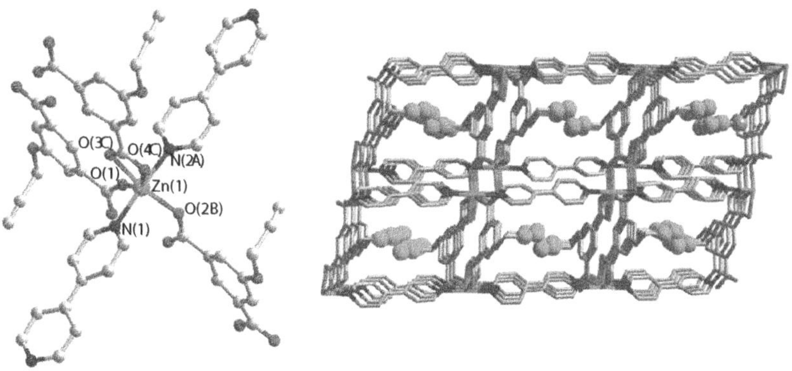

Figure 7.6 Asymmetry unit in Zn-DPYI crystal structure (left). Crystal structure of Zn-DPYI (right) [26].

are exposed within one dimensional channels (Figure 7.6, right). Of course, analyses such as elemental analysis and PXRD showed that the produced bulky sample has pure phase.

The prepared chiral MOFs post-synthetically were also investigated by CD analysis. The presence of D-AMP and L-AMP in the Zn-MOFs structures exhibited Cotton effects in spectrum. Most chiral materials have been synthesized by self-assembly having enantiopure organic linkers and metal ions, so post-synthesis modification method based on the used strategy in this work can be considered as effective method. According to frameworks nature, asymmetric aldol reaction was studied between aromatic aldehydes and cyclohexanone (Table 7.2). The control experiments were investigated such as using unmodified Zn-DPYI, chiral agent as homogeneous catalyst and Zn-MOF1 with four substrates 4-Nitrophenyl, 3-Nitrophenyl,

Table 7.2 Aldol reactions catalyzed by Zn-MOF1.[a]

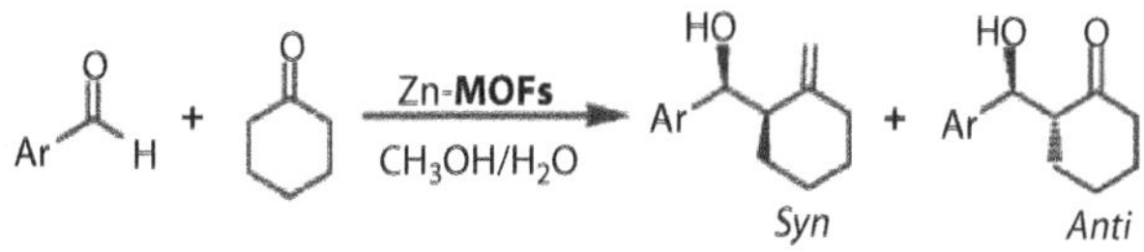

Entry	Ar	Catalyst	Yield (%)[b]	ee (%)[c]
1	4-Nitrophenyl	**Zn-MOF1**	75	70
		Zn-DPYI	<10	n.d.
		L-AMP	40	26
2	3-Nitrophenyl	**Zn-MOF1**	45	65
		Zn-DPYI	12	n.d.
		L-AMP	45	29
3	2-Nitrophenyl	**Zn-MOF1**	22	48
		Zn-DPYI	<10	n.d.
		L-AMP	17	<10
4	3-Formyl-1-phenylene(3,5-di-*tert*-butylbenzoate)	**Zn-MOF1**	Trace	n.d.
		Zn-DPYI	Trace	n.d.
		L-AMP	24	n.d.

[a]Reactions conditions: 25°C for 7 days in 1:1 solution of methanol and water using 0.5 mmol aldehyde and 5 mmol of cyclohexanones with the catalyst about 0.045 mmol (9% mol) and co-catalyst HAC 0.07 mmol (14% mol).
[b]Isolated yield based on aldehydes.
[c]Value represents the anti-isomer; n.d., not determined.

2-Nitrophenyl, and 3,5-di-tert-butylbenzoate at room temperature that the obtained results have been shown. The catalytic experiments by comparing the results obtained from different systems of Zn-MOF1 and Zn-MOF2 were completed under the same catalytic reactions conditions in Table 7.2 (Table 7.3). As shown, in Table 7.3, these catalysts illustrated same activity in this kind of asymmetric catalytic reaction. So, it is resulted that design and asymmetric preparation are effective especially in biological field.

So, during the post-synthesis method, achiral MOFs can be transformed to chiral MOFs with capability of the vast range of asymmetric catalysis with suitable efficiency [30–37]. With entering the chiral species into MOF framework, MOF gets out from homogeneous model and heterogeneity is induced to system that it causes the structural complexity. Another route in post-chiral modification of MOFs is incorporation of chiral agent on metal node. The incorporation of chiral agent to metal node is as an indirect method. This kind of chiral post-modification is applied to chiralize achiral framework of MOFs. When achiral MOFs have unsaturated sites to coordinate chiral compound and or they have functional groups that can be modified by chiral agent. The chiralization of inside achiral parent MOFs is an important point, but surely, this process will be little because MOFs usually do not have suitable pores with big size. This method is an impressive, simple, and straight method for chiralization of surface and immobilization of active centers [38–41].

Table 7.3 Aldol reactions by using Zn-MOF1 and Zn-MOF2.

Entry	**Ar**	**Catalysts**	**Yield (%)**	**ee (%)**
1	4-Nitrophenyl	**Zn-MOF1**	75	70
		Zn-MOF2	76	–73
2	3-Nitrophenyl	**Zn-MOF1**	45	65
		Zn-MOF2	49	–64
3	2-Nitrophenyl	**Zn-MOF1**	22	48
		Zn-MOF2	19	–73

According to the kind of functional groups, size, shape in MOFs framework, complexity can be generated.

The surface modification depends on the type of the presence functions and open metal sites. For instance, newly, we have reported the synthesis of a CMOF through post-chiral modification on metal node with a chiral species (tartrate anion) (Scheme 7.1) [42]. This CMOF was from the kind of MILs that was synthesized due to the ion-exchange, and finally, it was used as a chiral hydroxyl-rich MOF catalyst in two kinds of asymmetric reactions: enantioselective epoxidation and methanolysis.

The synthesized chiral MOF in this work was based on achiral MIL-101(Cr) that through its Cr nodes could be chiralized by using tart anion. As shown in Figure 7.7, there are separate catalytic centers, Brønsted and Lewis acids. Cr open metal sites are generated from removing adsorbed solvents under activation process. During this route, MIL framework is not damaged. So, Cr as Lewis acid centers acted in the enantioselective epoxidation. The presence of Brønsted acid centers in structure, tartrate anions, caused that this catalyst could show ability in asymmetric methanolysis. Authors claimed to chiralize a non-functionalized MOF as an acidic heterogeneous catalyst with mixing catalytic sites. Their

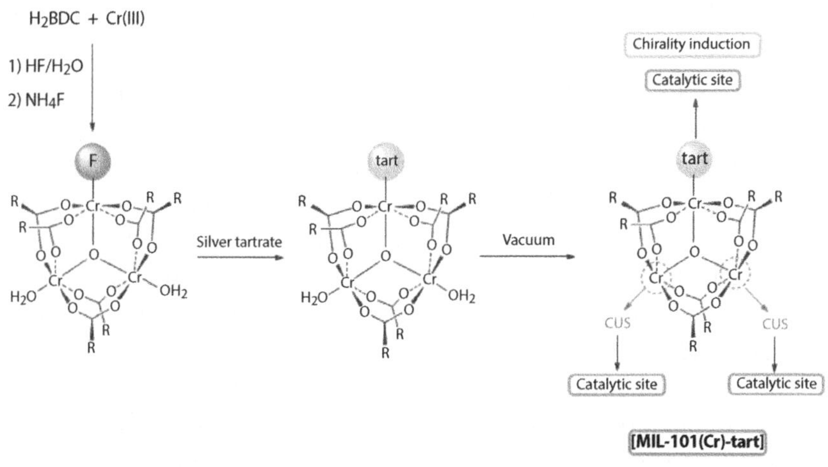

Scheme 7.1 Synthesis method of chiral [MIL-101(Cr)-tart].

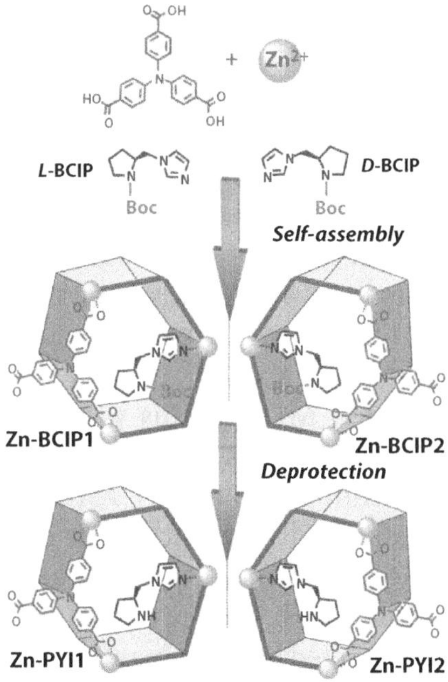

Figure 7.7 The preparation steps of one kind of chiral Zn-MOF through post-synthesis modification by using two chiral compounds with the opposite chirality [45].

suggested preparation process is based on the specific interactions, as follows:

There is an electrostatic attraction in silver tartrate salt that has been prepared from L-(+)-tartaric acid and Silver nitrate. In the first step, for conversion of neutral MIL-101(Cr) into chiral framework, the stripping of F anions was done by silver tartrate. This method advantage is employing a simple chiral species to eliminate the primitive anions in structure. The non-covalent acid-base interaction between Ag and F goes toward preparation of CMIL (Scheme 7.2).

Until now, it has been reported which such MOFs can be employed as ion exchangers due to their nature: cation or anion exchangers. In Table 7.4, control experiments for enantioselective epoxidation of goal substrate, styrene, have been showed. After determination of

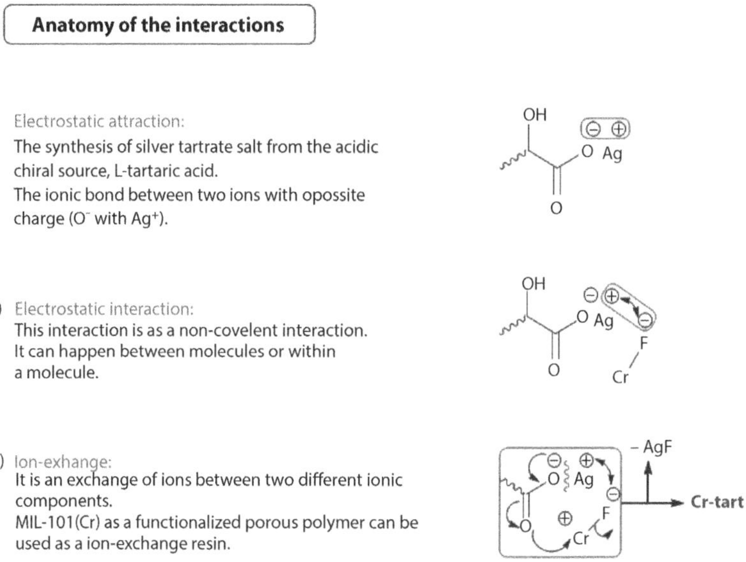

Scheme 7.2 Interactions in the preparation of [MIL-101(Cr)-tart].

optimal catalytic conditions, enantioselective epoxidation of several olefins and methanolysis of styrene oxide have been also investigated (Table 7.5). The findings show that not only framework provides catalytic sites but also its manipulation with different functional groups can prepare new design and structure. The experimental results showed that in addition to the metal sites as Lewis acid centers, the abundant and accessible of chiral functions can show catalytic activity as a Brønsted acid with high enantiomeric excess. These kinds of MOFs are rare so their synthesis is valuable [43].

Lately, CMOFs were synthesized through post-synthetic modification by the different chiral species from diverse families as phenylpropanoids, derivative of camphor and tartaric acid, amino acid, and asymmetric anhydrides [19, 25]. Chiral modification of MOFs can induce heterogeneity into frameworks and provide positions to interaction with diverse chiral molecules during the asymmetric processes. The important point is that the functional groups in frameworks, on linker and or node, can be functionalized by chiral compounds *via* postsynthetic modification so they can affect the chemical complexity [19].

Table 7.4 Control reactions for asymmetric epoxidation of styrene.[a]

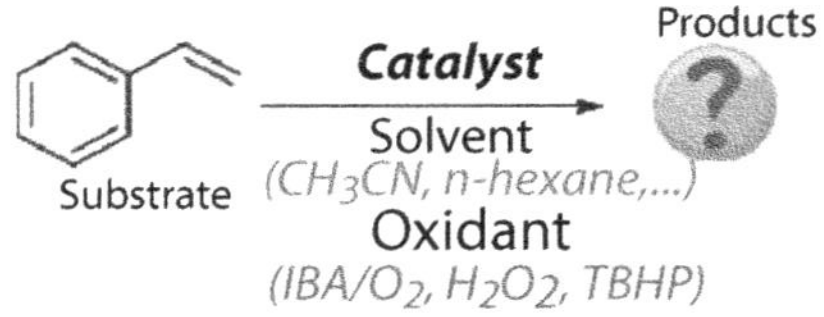

Entry	Catalyst	Catalytic conditions	Conv. (%)[b]	Epoxide selectivity (%)	Ee (%) (Conf.)[c]
1	***[MIL-101(Cr)-tart]***	**IBA/O_2/CH_3CN/80°C/8h**	**87**	**70**	***89 (S)***
2	*[MIL-101(Cr)-tart]*	H_2O_2/CH_3CN/80°C/6h	100	16	*68 (S)*
3	*[MIL-101(Cr)-tart]*	TBHP/CH_3CN/80°C/8h	46	81	*65 (S)*
4	*[MIL-101(Cr)-tart]*	No co-catalyst/O_2/ CH_3CN/80°C/8h	N.R.		
5	*[MIL-101(Cr)-tart]*	IBA/O_2/n-hexane/80°C/8h	28	23	*44 (S)*
6	*[MIL-101(Cr)-tart]*	IBA/O_2/EtOAc/80°C/8h	49	37	*61 (S)*

(Continued)

Table 7.4 Control reactions for asymmetric epoxidation of styrene.[a] (*Continued*)

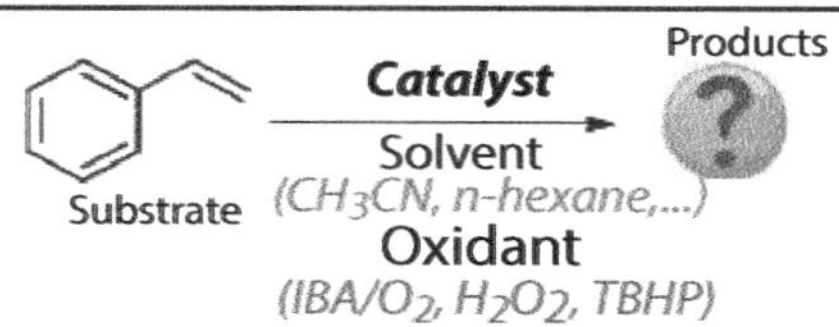

Entry	Catalyst	Catalytic conditions	Conv. (%)[b]	Epoxide selectivity (%)	Ee (%) (Conf.)[c]
7	*[MIL-101(Cr)-tart]*	IBA/O_2/CH_3OH/80°C/8h	6	17	*34 (S)*
8	*[MIL-101(Cr)] without L-tartrate ion*	IBA/O_2/CH_3CN/80°C/8h	79	68	
9	*[MIL-101(Cr)-tart]*	IBA/O_2/CH_3CN/40°C/8h	42	24	*61 (S)*

[a]Reaction conditions: catalyst 50.0 mg, ratio of ^{i}PrCHO/styrene: 3 mmol, solvents (CH_3CN,DMF, Toluene, and CH_3OH) 5 ml, molecular oxygen1 atm, 80°C.

[b]Conversions are based on the starting substrate and determined by GC.

[c]Enantiomeric excess (Ee%) was determined by GC on a chiral SGE-CYDEX-B capillary column. H_2O_2 6 mmol. TBHP 6 mmol.

Table 7.5 Asymmetric epoxidation of several olefins and methanolysis of styrene oxide.[a]

Entry	*Substrate*	Conversion (%)[b]/Time (h)	Epoxide selectivity (%)	Ee (%) (Conf.)[c]
1	*α-methyl styrene*	83/8	68	*85 (S)*
2	*1-Phenyl-1-cyclohexene*	79/8	100	*78 (R,R)*
3	*1-Octene*	77/8	100	*100 (R or S)*
4	*1-Decene*	69/8	100	*91 (R)*
5	*trans-stilbene*	100/30 min	100	*75 (R,R)*

[a]Reaction conditions: catalyst 50.0 mg (0.0013 mmol Cr/g catalyst), ^{i}PrCHO/substrate 3 mmol, CH_3CN 5 ml, molecular oxygen 1 atm.
[b]Conversions are based on the starting substrate and determined by GC.
[c]Enantiomeric excess (Ee%) was determined by GC on a chiral SGE-CYDEX-B capillary column. H_2O_2 6 mmol. TBHP 6 mmol.

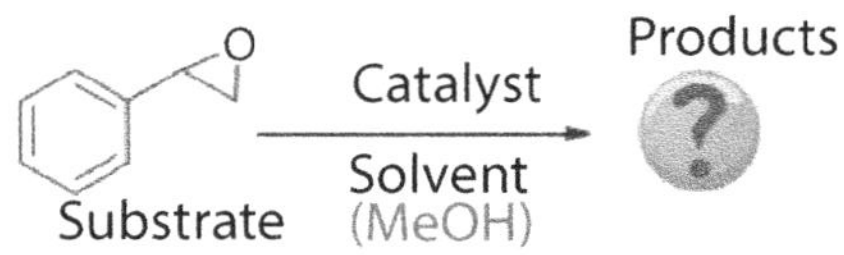

Entry	*Catalytic center*	*Conversion (%)[b]/ Time (h)*	*2-methoxy-2-phenylethanol Selectivity(%)*	*Ee (%) (Conf.)[c]*
1	*Cr^{3+} (Lewis acid)*	10/48	100	
2	***[MIL-101(Cr)-tart]*** *(Contains Brønsted and Lewis acid)*	**100/9**	**100**	***90 (S)***

[a]Reaction conditions: catalyst 50.0 mg, styrene oxide 1 mmol, MeOH 10 ml, 40°C.
[b]Conversions are based on the starting substrate and determined by GC.
[c]Enantiomeric excess (Ee%) was determined by GC on a chiral SGE-CYDEX-B capillary column. H_2O_2 6 mmol. TBHP 6 mmol.

If the used chiral species have big size, they will affect the order of MOFs architecture. Sometimes, CMOFs can be prepared *via* postsynthetic modification on nodes that are not open metal sites and they have functional groups. So, node functional groups can be functionalized with chiral species. Chiral NU-1000 is one kind of the chiral Zr-based MOFs that NU-1000 was functionalized by using tartaric acid through SALI method in it. This post-chiral modification can affect homogeneity of parent MOF (Scheme 7.3) [44]. In this work, a stable chiral MOF by using NU-1000 and chiral tartaric acid through was synthesized. Its preponderance was simple design and synthesis with high density along with of incorporated chiral species and Mo-catalysts as the Lewis acid sites. The simplified design is an ideal issue in preparing this CMOF. The enantioselective catalytic activity of [C-NU-1000-Mo] was investigated in the epoxidation of different prochiral alkenes to the considered epoxides with high enantiomeric excesses.

So, this chiral heterogeneous catalyst can be considered as a stable CMOF catalyst with excellent catalytic results without remarkable degradation in its activity. In Figure 7.7, another example of chiral MOF that was prepared by post-chiral synthesis has been shown [45].

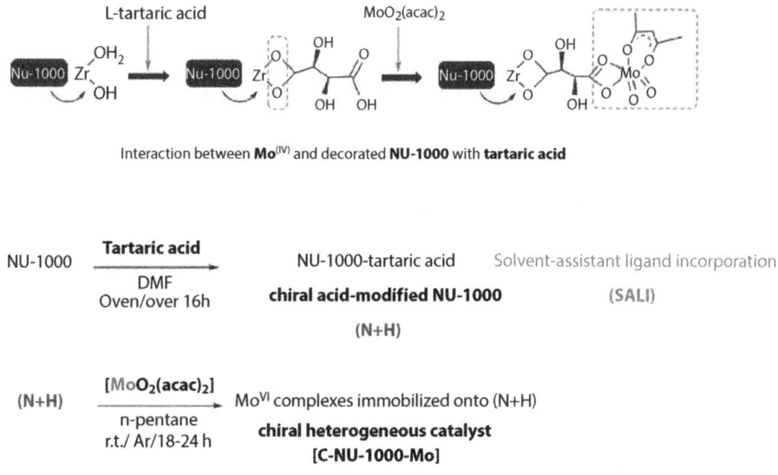

Scheme 7.3 The preparation steps of post-chiral modification NU-1000 by SALI method and then Mo-catalyst immobilization on chiral species.

7.3 Conclusion

With the increased interest in using MOFs in various applications, considerable efforts have been performed in the creation of chiral MOFs as chiral porous materials that are robust and stable. However, there is still difficult to achieve the creation of chiral MOFs. Chemical and physical stability, porosity, and capability of chemical manipulation of frameworks have attracted much more attention compared with the other materials. The effective chiralization strategies have been discussed with some of the examples but it should be noted that all of them were about pre-synthesis of chiral MOFs. In relation to asymmetric synthesis of MOFs; in this chapter, two kinds of strategies were introduced with several examples: chiral-template synthesis and post-chiral modification. In post-synthesis strategy, correct selection of parent MOF is very important for conversion of achiral framework to its chiral counterpart. In this kind of MOFs, the presence of functional groups or open metal sites in structure are main factors. Indeed, post-synthetic modification can be applied to either node or ligands or even the pores. So, size and shape of the pores should be well in this way. With this method, the vast range of chiral MOFs can be constructed with diverse chemistry through different kinds of attachments with various applications that some of them were new. Using chiral template is another effective synthetic method for the induction of chiral centers into MOFs framework. Of course, the use of achiral template can produce chiral centers in structures, too. Selection of template molecule is the key in this strategy. Ionic liquids and L-proline can be effective templates to obtain chiral MOFs. They can induce distortion into MOFs skeleton. The main issue of this strategy is that template selection is not generally applicable, which makes it very challenging in the future investigations of chiral MOFs.

References

1. Kepert, C., Prior, T., Rosseinsky, M.J., A versatile family of interconvertible microporous chiral molecular frameworks: the first example

of ligand control of network chirality. *J. Am. Chem. Soc.*, *122*, 21, 5158–5168, 2000.
2. Dang, D., Wu, P., He, C., Xie, Z., Duan, C., Homochiral metal–organic frameworks for heterogeneous asymmetric catalysis. *J. Am. Chem. Soc.*, *132*, 41, 14321–14323, 2010.
3. Zhang, J., Chen, S., Wu, T., Feng, P., Bu, X., Homochiral crystallization of microporous framework materials from achiral precursors by chiral catalysis. *J. Am. Chem. Soc.*, *130*, 39, 12882–12883, 2008.
4. Zhang, S.-Y., Li, D., Guo, D., Zhang, H., Shi, W., Cheng, P., Wojtas, L., Zaworotko, M.J., Synthesis of a chiral crystal form of MOF-5, CMOF-5, by chiral induction. *J. Am. Chem. Soc.*, *137*, 49, 15406–15409, 2015.
5. Han, Q., He, C., Zhao, M., Qi, B., Niu, J., Duan, C., Engineering chiral polyoxometalate hybrid metal–organic frameworks for asymmetric dihydroxylation of olefins. *J. Am. Chem. Soc.*, *135*, 28, 10186–10189, 2013.
6. Han, Y.-H., Liu, Y.-C., Xing, X.-S., Tian, C.-B., Lin, P., Du, S.-W., Chiral template induced homochiral MOFs built from achiral components: SHG enhancement and enantioselective sensing of chiral alkamines by ion-exchange. *Chem. Commun.*, *51*, 77, 14481–14484, 2015.
7. Ernst, S., Fuchs, E., Yang, X., Materials, M., Enantioselective hydrogenation on zeolite-encapsulated chiral palladium–salen complexes. *Microporous Mesoporous Mater.*, *35*, 137–142, 2000.
8. Sabater, M.J., Corma, A., Domenech, A., Fornés, V., García, H., Chiral salen manganese complex encapsulated within zeolite Y: a heterogeneous enantioselective catalyst for the epoxidation ofalkenes. *Chem. Commun.*, 14, 1285–1286, 1997.
9. Ogunwumi, S.B. and Bein, T., Intrazeolite assembly of a chiral manganese salen epoxidation catalyst. *Chem. Commun.*, 9, 901–902, 1997.
10. Horcajada, P., Salles, F., Wuttke, S., Devic, T., Heurtaux, D., Maurin, G., Vimont, A., Daturi, M., David, O., Magnier, E., How linker's modification controls swelling properties of highly flexible iron (III) dicarboxylates MIL-88. *J. Am. Chem. Soc.*, *133*, 44, 17839–17847, 2011.
11. Morris, R.E. and Bu, X., Induction of chiral porous solids containing only achiral building blocks. *Nat. Chem.*, *2*, 5, 353, 2010.
12. Jing, X., He, C., Dong, D., Yang, L., Duan, C., Homochiral crystallization of metal–organic silver frameworks: asymmetric [3+ 2] cycloaddition of an azomethine ylide. *Angew. Chem. Int. Ed.*, *51*, 40, 10127–10131, 2012.
13. Wen, Y., Sheng, T., Sun, Z., Xue, Z., Wang, Y., Wang, Y., Hu, S., Ma, X., Wu, X., A combination of the "pillaring" strategy and chiral induction: an approach to prepare homochiral three-dimensional coordination

polymers from achiral precursors. *Chem. Commun.*, *50*, 61, 8320–8323, 2014.

14. Livage, C., Guillou, N., Rabu, P., Pattison, P., Marrot, J., Férey, G., Bulk homochirality of a 3-D inorganic framework: ligand control of inorganic network chirality. *Chem. Commun.*, 30, 4551–4553, 2009.
15. Wang, Z. and Cohen, S.M., Postsynthetic modification of metal–organic frameworks. *Chem. Soc. Rev.*, *38*, 5, 1315–1329, 2009.
16. Bisht, K.K., Parmar, B., Rachuri, Y., Kathalikattil, A.C., Suresh, E., Progress in the synthetic and functional aspects of chiral metal–organic frameworks. *CrystEngComm*, *17*, 29, 5341–5356, 2015.
17. Bernt, S., Guillerm, V., Serre, C., Stock, N., Direct covalent post-synthetic chemical modification of Cr-MIL-101 using nitrating acid. *Chem. Commun.*, *47*, 10, 2838–2840, 2011.
18. Jiang, J.-Q., Yang, C.-X., Yan, X.-P., Postsynthetic ligand exchange for the synthesis of benzotriazole-containing zeolitic imidazolate framework. *Chem. Commun.*, *51*, 30, 6540–6543, 2015.
19. Garibay, S.J., Wang, Z., Tanabe, K.K., Cohen, S.M., Postsynthetic modification: a versatile approach toward multifunctional metal-organic frameworks. *Inorg. Chem.*, *48*, 15, 7341–7349, 2009.
20. Tanabe, K.K. and Cohen, S.M., Engineering a metal–organic framework catalyst by using postsynthetic modification. *Angew. Chem. Int. Ed.*, *48*, 40, 7424–7427, 2009.
21. Du, Y., Li, X., Lv, X., Jia, Q., interfaces, Highly Sensitive and Selective Sensing of Free Bilirubin Using Metal–Organic Frameworks-Based Energy Transfer Process. *ACS Appl. Mater. Interfaces*, *9*, 36, 30925–30932, 2017.
22. Zhu, H., Wang, L., Jie, X., Liu, D., Cao, Y., interfaces, Improved interfacial affinity and CO_2 separation performance of asymmetric mixed matrix membranes by incorporating postmodified MIL-53 (Al). *ACS Appl. Mater. Interfaces*, *8*, 34, 22696–22704, 2016.
23. Garibay, S.J., Wang, Z., Tanabe, K.K., Cohen, S.M., Postsynthetic modification: a versatile approach toward multifunctional metal-organic frameworks. *Inorg. Chem.*, *48*, 15, 7341–7349, 2009.
24. Serra-Crespo, P., Ramos-Fernandez, E.V., Gascon, J., Kapteijn, F., Synthesis and characterization of an amino functionalized MIL-101 (Al): separation and catalytic properties. *Chem. Mater.*, *23*, 10, 2565–2572, 2011.
25. Kou, W.-T., Yang, C.-X., Yan, X.-P., Post-synthetic modification of metal–organic frameworks for chiral gas chromatography. *J. Mater. Chem. A*, *6*, 37, 17861–17866, 2018.

26. Zhu, W., He, C., Wu, P., Wu, X., Duan, C., "Click" post-synthetic modification of metal-organic frameworks with chiral functional adduct for heterogeneous asymmetric catalysis. *Dalton Trans.*, *41*, 10, 3072–3077, 2012.
27. Trost, B.M. and Brindle, C.S., The direct catalytic asymmetric aldol reaction. *Chem. Soc. Rev.*, *39*, 5, 1600–1632, 2010.
28. Gruttadauria, M., Giacalone, F., Noto, R., Supported proline and proline-derivatives as recyclable organocatalysts. *Chem. Soc. Rev.*, *37*, 8, 1666–1688, 2008.
29. Guillena, G., Hita, M. d. C., Najera, C., Viozquez, S.F., A Highly Efficient Solvent-Free Asymmetric Direct Aldol Reaction Organocatalyzed by Recoverable (S)-Binam-l-Prolinamides. ESI-MS Evidence of the Enamine–Iminium Formation. *J. Org. Chem.*, *73*, 15, 5933–5943, 2008.
30. Sharpless, K.B., Kolb, H., Finn, M., Click chemistry: diverse chemical function from a few good reactions. *Angew. Chem. Int. Ed.*, *40*, 2004–2021, 2001.
31. Rostovtsev, V., Green, L., Fokin, V., Sharpless, K., Kolb, H., Finn, M., Click chemistry: diverse chemical function from a few good reactions. *Angew. Chem. Int. Ed.*, *41*, 2596–2599, 2002.
32. Helms, B., Mynar, J.L., Hawker, C.J., Frechet, J.M., Dendronized linear polymers via "click chemistry". *J. Am. Chem. Soc.*, *126*, 46, 15020–15021, 2004.
33. Soriano del Amo, D., Wang, W., Jiang, H., Besanceney, C., Yan, A.C., Levy, M., Liu, Y., Marlow, F.L., Wu, P., Biocompatible copper (I) catalysts for *in vivo* imaging of glycans. *J. Am. Chem. Soc.*, *132*, 47, 16893–16899, 2010.
34. Moses, J.E. and Moorhouse, A.D., The growing applications of click chemistry. *Chem. Soc. Rev.*, *36*, 8, 1249–1262, 2007.
35. Goto, Y., Sato, H., Shinkai, S., Sada, K., "Clickable" metal–organic framework. *J. Am. Chem. Soc.*, *130*, 44, 14354–14355, 2008.
36. Savonnet, M., Bazer-Bachi, D., Bats, N., Perez-Pellitero, J., Jeanneau, E., Lecocq, V., Pinel, C., Farrusseng, D., Generic post functionalization route from amino-derived metal–organic frameworks. *J. Am. Chem. Soc.*, *132*, 13, 4518–4519, 2010.
37. Zhao, D., Tan, S., Yuan, D., Lu, W., Rezenom, Y.H., Jiang, H., Wang, L.Q., Zhou, H.C., Surface functionalization of porous coordination nanocages via click chemistry and their application in drug delivery. *Adv. Mater.*, *23*, 1, 90–93, 2011.
38. Rieter, W.J., Taylor, K.M., Lin, W., Surface modification and functionalization of nanoscale metal-organic frameworks for controlled

release and luminescence sensing. *J. Am. Chem. Soc.*, *129*, 32, 9852–9853, 2007.

39. Burrows, A.D., Mixed-component metal–organic frameworks (MC-MOFs): enhancing functionality through solid solution formation and surface modifications. *CrystEngComm*, *13*, 11, 3623–3642, 2011.
40. Cohen, S.M., Modifying MOFs: new chemistry, new materials. *Chem. Sci.*, *1*, 1, 32–36, 2010.
41. Shi, W., Cao, L., Zhang, H., Zhou, X., An, B., Lin, Z., Dai, R., Li, J., Wang, C., Lin, W., Surface Modification of Two-Dimensional Metal–Organic Layers Creates Biomimetic Catalytic Microenvironments for Selective Oxidation. *Angew. Chem. Int. Ed.*, *56*, 33, 9704–9709, 2017.
42. Berijani, K. and Morsali, A., Dual activity of durable chiral hydroxyl-rich MOF for asymmetric catalytic reactions. *J. Catal.*, *378*, 28–35, 2019.
43. Li, B., Leng, K., Zhang, Y., Dynes, J.J., Wang, J., Hu, Y., Ma, D., Shi, Z., Zhu, L., Zhang, D., Metal–organic framework based upon the synergy of a Brønsted acid framework and Lewis acid centers as a highly efficient heterogeneous catalyst for fixed-bed reactions. *J. Am. Chem. Soc.*, *137*, 12, 4243–4248, 2015.
44. Berijani, K., Morsali, A., Hupp, J.T., An Effective Strategy for Creating Asymmetric MOFs for Chirality Induction: A Chiral Zr-based MOF for Enantioselective Epoxidation. *Catal. Sci. Technol.*, *9*, 3388–3397, 2019.
45. Wu, P., He, C., Wang, J., Peng, X., Li, X., An, Y., Duan, C., Photoactive Chiral Metal-Organic Frameworks for Light-Driven Asymmetric α-Alkylation of Aldehydes. *J. Am. Chem. Soc.*, *134*, 36, 14991–14999, 2012.
46. Han, Q., Qi, B., Ren, W., He, C., Niu, J. and Duan, C., Polyoxometalate-based homochiral metal-organic frameworks for tandem asymmetric transformation of cyclic carbonates from olefins. *Nat. Commun.*, *6*, 1, 1-8, 2015.

8
Complexity Based on Structural Defects

Abstract
"Defect" in materials is an exciting concept that can be also appeared in metal-organic frameworks. This property can be produced inherently or intentionally and it can provide new opportunities in materials for different applications. Different kinds of defects can produce heterogeneous structures with various degrees of complexity and chemical-physical properties. So, in this chapter, we discuss the generated heterogeneity from structural disorders and defects in MOFs and its effects. It should be said that defects do not necessarily have adverse effects.

Keywords: Defect engineering, structural disorder, heterogeneity, crystal irregularities, inherent defects, heterostructure

8.1 Inherent Defect

Until now, various MOFs have been designed and synthesized. The structural defects can be considered as disorders that can exit the atoms from the regular arrangement and create heterogeneity in the structure. Usually, the complexity and defects are not seen only in MOFs but they can be observed in the other materials and they may affect properties and applications [1–8].

As well, imperfections are in solids even in the crystals. It is ideal that the matter to be perfect, for example, Colin John Humphreys is the former Goldsmiths' professor of materials science, who says: "Crystals are like people, it is the defect in them which tends to make them interesting!". This sentence needs much thinking. Recently, study on the complex nature of this phenomenon has attracted attention. In many articles, these phrases have been brought: "Inherent

Ali Morsali and Kayhaneh Berijani. Metal-Organic Frameworks with Heterogeneous Structures, (149–170)
© 2021 Scrivener Publishing LLC

defect and Designed defect". Even, one of sections in the review by Yaghi *et al.* that has been published for 150 years antiquity of BASF is "MOFs with Random and Ordered Defects". They presented this topic in several examples with the various defects strategies in MOFs in which these instances mostly included fault engineering that were generated intentionally. Some researchers investigated the various kinds of defects in MOFs and published several papers with total subject that some of them have been mentioned in Table 8.1. In Figure 8.1, a plot has been shown based on the number of published articles about defect of MOFs. Our aim, from the discussion about this section, is to present more information in comparison to the past according to MOFs that have substantial structural irregularities.

Many developments have been observed about design and synthesis of ideal MOFs but there are not defect-free crystalline MOFs [22]. Most of them have defects that ultimately lead to diversity but there are methods that decrease and control MOFs imperfection. Defects are one of the reasons to produce heterogeneity in compounds and inherent defect can be produced in different levels with changes in the preparing conditions of crystals of MOFs [23].

In this chapter, we report defects that are produced instinctively during synthesis. As we know, MOFs, as porous materials and crystalline, have different defects which are totally two states. These defects are formed intrinsically during the synthesis due to stacking fault and even partial dislocations without handworks [14]. Inherent defect can be to different forms like the produced vacancies due to the missing linkers or nodes, the presence of different pore volumes, and the plausible collapse and interpenetration of MOFs frameworks [24–27]. Sometimes, the impurities of the starting compounds can enter into framework and generate defects. The important point is that the impurities are not intentionally added during synthesis. In the synthesis of defective MOFs, inherent defects can be also formed through misconnections or dislocations in different models during crystallization process or post-crystallization cleavage [28, 29]. Dislocations that are generated in MOFs crystalline structure can be screwed due to the growth spirals [28, 30]. These kinds of defects are known as electrical faults that are used in optical and conduction field. The structural and proton-conductive properties that are

Table 8.1 Some of the published papers about defects in metal-organic frameworks.

No.	Authors	Title of articles	Year	Ref.
1	Heine *et al.*	Defects in MOFs: A Thorough Characterization	2012	[9]
2	Roeffaers *et al.*	Three-Dimensional Visualization of Defects Formed during the Synthesis of Metal–Organic Frameworks: A Fluorescence Microscopy Study	2013	[10]
3	Goodwin *et al.*	Structural Disorder in Molecular Framework Materials	2013	[11]
4	Goodwin *et al.*	Correlated Defect Nanoregions in A Metal–Organic Framework	2014	[12]
5	Lively *et al.*	Defects in Metal–Organic Frameworks: Challenge or Opportunity?	2015	[13]
6	Fischer *et al.*	Defect-Engineered Metal–Organic Frameworks	2015	[14]
7	Goodwin *et al.*	Defects and Disorder in Metal–Organic Frameworks	2016	[15]
8	Lillerud *et al.*	Defect Engineering: Tuning the Porosity and Composition of the Metal–Organic Framework UiO-66 via Modulated Synthesis	2016	[16]
9	Coudert *et al.*	Interplay Between Defects, Disorder and Flexibility in Metal–Organic Frameworks	2017	[17]

(*Continued*)

Table 8.1 Some of the published papers about defects in metal-organic frameworks. (*Continued*)

No.	Authors	Title of articles	Year	Ref.
10	Ren *et al.*	Structural Defects in Metal–Organic Frameworks (MOFs): Formation, Detection and Control Towards Practices of Interests	2017	[18]
11	Fischer *et al.*	Defective Metal–Organic Frameworks	2018	[19]
12	Shi *et al.*	Defect Engineering in Metal–Organic Frameworks: A New Strategy to Develop Applicable Actinide Sorbents	2018	[20]
13	Kim *et al.*	Defect Engineering into Metal–Organic Frameworks for the Rapid and Sequential Installation of Functionalities	2018	[21]

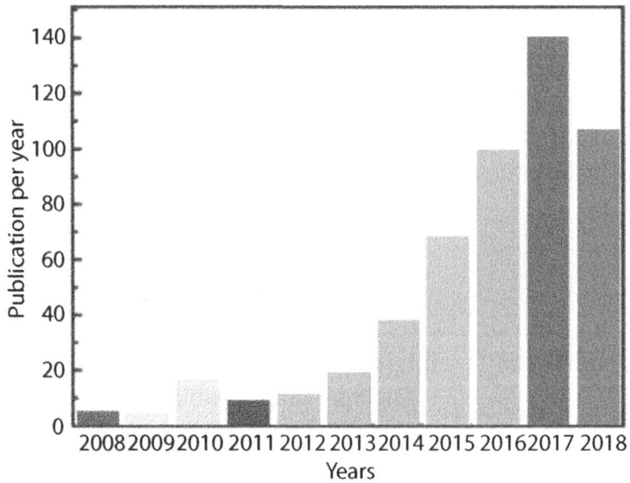

Figure 8.1 A plot based on the number of published articles about the defect of MOFs.

interesting features were investigated in some of the materials with imperfections, too [31].

Some of the defects as internal faults in MOFs structure appear when organic linkers are pendant and or missing organic linkers can happen. The synthesis of this kind of MOFs is related to the preparing conditions as temperature and the ratio of the used components [32–35]. This missing linker defect in MOFs can also happen according to the interconnected cluster nodes by organic linkers. In this model, the inherent defect can be probably produced due to the missing linkers that can be one or more. The missing nodes can happen during the growth of crystal [14]. The rate of crystal growth is also effective in the creation of the vacancy in MOFs because there is enough time to put the components in sufficient position in framework [32]. In the frameworks with interpenetrations, defect can be observed due to the generated activating binding sites (Figure 8.2) [11, 36].

Sometimes, the defects, based on the position of their vacancies, can be "Ordered defects" and certainly they have different effects on properties and applications of materials [37]. The effect of temperature changes was investigated on the geometrical and mechanical behavior of these MOFs. The obtained results showed that the generated defects can influence on the physical features of compounds, too [38].

Interestingly, the inherent defect can also occur in SURMOFs due to the different factors. During the optimization of the orientation growth in SURMOFs, the organic linker concentrations have an important role in the generation of defect [39–41]. A simple method

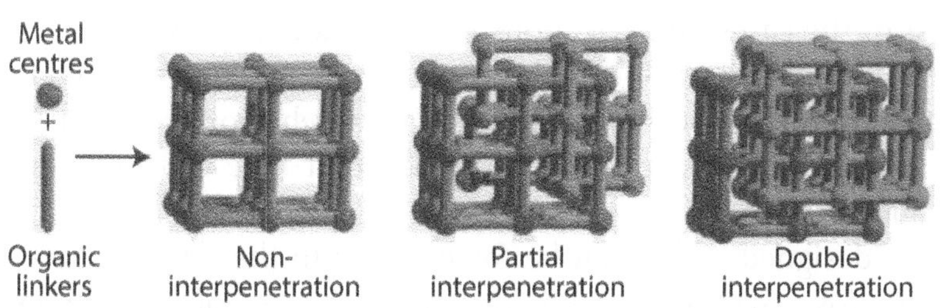

Figure 8.2 MOF with interpenetration [76].

to analysis of the available defects is thermogravimetric analysis. The other technique to identify defects level is imaging based on fluorescence microscopy specifically in the scaffolds that have incorporated molecules as dye due to the present defects [23]. So, imaging shows both of the diversity degree in structure and defect level in framework. It is important that happened defects can affect the MOFs applications and improve their abilities as adsorption processes and catalytic activity [42–44]. Since, in the defective framework, the pore volume increases and it leads to suitable catalytic activity as a heterogeneous catalyst [45]. In the inherent defect, faults can be randomly created in a crystalline MOF. A kind of this defect was observed in PCN-160 with Zr and AZDC components that its preparation was done by solvothermal reaction which the linker exchange happened with using CBAB linker at suitable temperature. This process led to produce of random defects (missing linker) [46]. The distributed defects can be identified and assessed with using mass-nuclear magnetic and electron paramagnetic resonance spectroscopies as powerful techniques [47]. It seems that the defects of MOFs can be designed and controlled with the different methodologies. This type of defect has been known as "defect engineering" or "designed defect" that has been discussed in the next section.

8.2 Designed Defect

Another kind of defect is intentional and induces heterogeneity to structure named "defect engineering" [14]. The fabrication of designed defect is very important because during the manipulation of crystal structure, the quality of crystal should be remained. In most of the cases, imperfections can generate the new and specific features in materials. Surely, the kinds of defects, densities, and their distribution are not ineffective on MOFs behavior particularly on the complexity of structure that is different type of heterogeneity [48–50].

The designed defect can be formed due to the several strategies as De Novo Synthesis that is divided to classes of coassembly of different organic ligands or fragments of ligands that is related to solid-solution method. In this approach, final MOFs based on mixed linkers can maintain topology that is related to the type of linkers. In mixed ligand MOFs, mixed organic linkers can have differences about size, topology, and type that the obtained heterostructures can supply an individual level of heterogeneity. If MOFs have mixed linkers that are isostructure with different side groups, then the heterogeneity can be induced to frameworks. In Figure 8.3, the different cases of MOFs were compared to the parent MOF (Figure 8.3a) and their difference in this kind of MOFs is clear [51].

Another method that is not distinct from missing linker is templating procedure (soft and hard templates) that can be employed to generate defect engineering that leads to crystalline MOF with larger pores or the mixture of pores [52]. For example, in addition to hard templates, MOFs' instability under certain conditions can be used to generate defects and construct larger pores. In 2015, Mao *et al.* reported a ligand-assisted etching strategy without templates to synthesize mesoporous crystals and HKUST-1 membranes. In this work, pre-prepared MOF and membranes

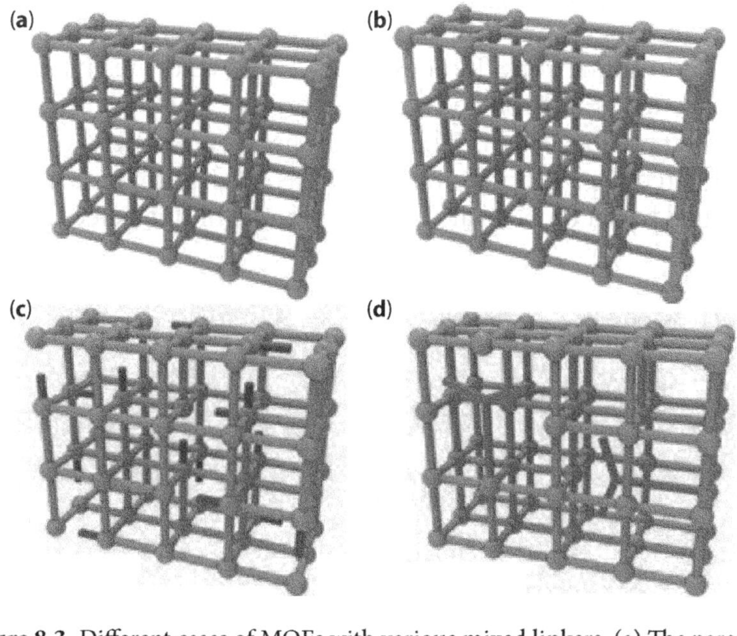

Figure 8.3 Different cases of MOFs with various mixed linkers. (a) The parent MOF, (b) the different mixed linker structurally, and (c and d) the large and truncated mixed organic linkers [51].

were immersed in specific solutions of H_3BTC with particular concentrations, following which H^+ ions from deprotonated H3BTC etched crystals to induce mesopores. Under specific conditions, meso-HKUST-1 crystals were synthesized with a BET surface area 1,462 m^2/g. Macroporous products are produced at more elevated temperatures (such as 80°C).

The designed defects can also occur in nodes that lead to vacancy sites of metal nodes in MOFs. In this case, metal atom can disappear from location so vacancy is created, although this defect is point defect that is observed in intrinsic defects. This type of fault can be seen along with linker vacancy, simultaneously. The design of imperfect nodes can be controlled by thermal methods [53, 54]. It can be said that temperature affects metal sites, and defects will exist in the lattice of MOFs [40]. Post-synthetic strategy is a suitable treatment to form this kind of defect that is classified to four subgroups: a) acid/base, b) SALI/SALE, and c) washing and activation. The appellation of this method is logical, because, here, the defect agents are added into the framework of MOFs after the process of preparation. The postsynthetic modification of parent MOF can happen by adding components as strong acids that are added into framework. In addition to Lewis acid sites in the parent structure, the Brønsted acid centers are also observed without any crystallinity losing [55]. This change in framework can be investigated *via* experimental test as gas adsorption and it can be identified based on porosity. Another model of this defect is SALE

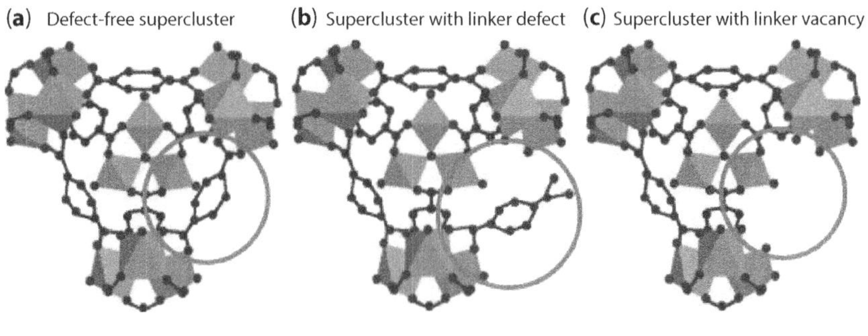

Figure 8.4 MIL to different forms: (a) without any defect, (b) dangling organic linker, and (c) linker vacancy. The orange polyhedral: cationic units. C atom: black. O atom: blue [60].

Table 8.2 Some of defective MOF-based materials.

Entry	Defective MOF-based materials	Defect type	Notice	Ref.
1	HKUST-1	Plane dislocations Fractures propagating	Catalytic activity	[64]
2	HKUST-1	Dislocation growth spirals	Hindered diffusion across dislocation core	[65]
3	HKUST-1	Missing carboxylates	Multiple N2 adsorbed at defects	[66]
4	HKUST-1	Temporary defects	High catalytic activity Selectivity in Knoevenagel reactions	[67]
5	HKUST-1 SURMOF	Linker vacancies	Increased reactivity towards CO	[68]
6	HKUST-1 with fragmented linkers	Partial metal node reduction Linker vacancies	Enhanced catalytic activity Altered porosity Selectivity in hydroxylation of toluene	[69]
7	HKUST-1 and NU-125 with fragmented linkers	Missing paddlewheel clusters Linker vacancies	Altered porosity H_2 and CH_4 uptake Functionalization of pore	[70]

(*Continued*)

Table 8.2 Some of defective MOF-based materials. (*Continued*)

Entry	Defective MOF-based materials	Defect type	Notice	Ref.
8	PCN-125 (linker-fragmentation)	Linker vacancies	Mesopore formation Increase in CO_2 uptake	[71]
9	MOF-5	Surface defects Cracks	Defects localized in a 10-mm shell	[72]
10	MOF-5 (microwave)	Surface defects Grooves	Increased synthesis time	[73]
11	MOF-5 and IRMOF-8	Lattice interpenetration	Enhanced H_2 uptake	[74]
12	UiO-66(Hf)	Linker and metal cluster Vacancies Correlated defect nanoregions	Altered porosity Altered mechanical properties	[75]
13	NOTT-202	Slits formed by defective	Temperature-dependent adsorption/ desorption hysteresis Selective response to CO_2	[76]
14	MOF-505	Metal vacancies	Long-range ferromagnetic coupling	[77]

(*Continued*)

Table 8.2 Some of defective MOF-based materials. (*Continued*)

Entry	Defective MOF-based materials	Defect type	Notice	Ref.
15	CAU-7 (Bi-BTB)	Lewis acidic defect CUSs	Weak Lewis acid catalyst for selective hydroxymethylation of 2-methylfuran	[78]
16	MIL-47	Linker vacancies Partial linker decoordination	Catalytic activity	[79]
17	ZIF-8 and ZIF-9	Various defect sites on surface	Catalytic activity	[80]
18	MOF-801-P	Linker vacancies	Defect-induced hydrophilicit	[81]

and SALI that are displacement and incorporation of intended organic linker in MOFs, respectively. During the first process, the replacement of the second linker with various structures can successfully happen with first organic linker along with high percent that can lead to pores openings with different size. So, the homogeneous system goes toward heterogeneous system and then defect is produced. SALE is a useful modification to generate MOFs with mixed linkers (Figure 8.4) [56–59].

This phenomenon has been also studied about zeolitic imidazolate frameworks [57]. In SALI method, there is the manipulation of node with the range of coordinated compound with the different functional groups as carboxylic acids that are coordinated to node site and increase the ability of framework through the acid and base interaction [61]. The considerable chemical reactions can be performed in the defective sites [42]. One of manipulations after temperature during the synthesis is a simple work: how to wash or activate that can generate conditions to produce defect [6]. However, the strong porous materials as Zr-based MOFs can maintain their framework, but in some of MOFs, the notable distortion can happen [62]. The imperfections can be also generated with different ions implantation such as protons, alpha, and boron by irradiation that can be studied by positron annihilation spectroscopy (PAS) [63]. Ion implantation defects have been widely studied about metals and alloys; therefore, we hope these kinds of defects are more evaluated in metal-organic frameworks. It is hard and complex but it can be done. Nothing is quite impossible! Some of these kinds of materials will be specifically discussed in Table 8.2.

Recently, Kim *et al.* reported UiO-66(Zr)-$(OH)_2$ framework with the structural defect for CO_2 adsorption. This MOF could be rapidly synthesized during 10 min by using zirconium chloride, 2,5-dihydroxylterephtalic acid, N,N-dimethylformamide, acetic acid and deionized (DI) water. After preparing process, the obtained product was dried over 100 °C (12 h).

The produced samples with chemical formula UiO-$(OH)_2$-x were generated. X parameter is molar ratio of CH_3COOH to $ZrCl_4$. Of course, under studied preparing conditions, it should be said that experiment without acetic acid was not performed.

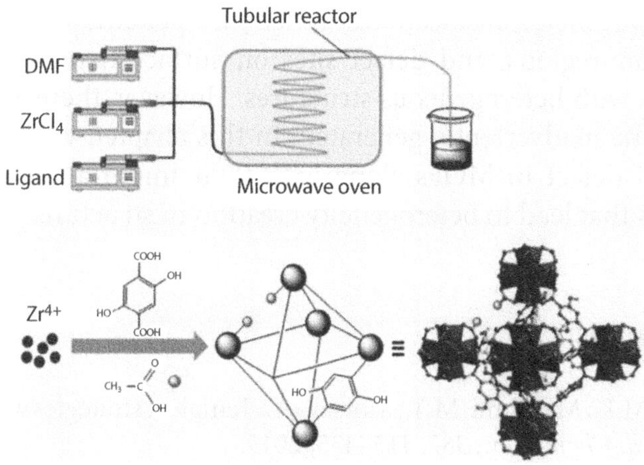

Figure 8.5 Defective UiO-66(Zr)-$(OH)_2$ framework for CO_2 capture [82].

Based on investigations, the exchange of bridging linker and the modulator in structure leads to defects production inside framework. Linker deficiency in framework increased to 0.60~2.30 per Zr_6 unit as molar ratio of acetic acid/Zr^{4+} increased to 250. Importantly, increased deficiency of ligand of per SBU led to a steadily increased adsorption capacity of CO_2 but decreasing CO_2 adsorption heat. These results propose that microwave-assisted continuous-flow synthesis can be suitable strategy to create defects in framework of MOF materials for different applications such as CO_2 capture (Figure 8.5) [82].

8.3 Conclusion

In nature, there are structural disorders or defects in solid-state materials and it should be said that they can strongly influence on physical and chemical properties of materials. In fact, defect is an effective tool that can produce materials with heterogeneity and heterogeneous structures. This property can induce the various capabilities and specific properties into MOFs framework for different applications such as catalysis, gas storage, separation, and electrical conductivity. So, recently, types of defects and their identification along with formation methods have been investigated like linker or node vacancies, dislocations, lattice interpenetration, correlated defect nanoregions, and defect sites on surface that can produce materials with heterogeneous structures. However, there are defects that can be inadvertently generated. In this chapter, we introduced details of defect in MOFs along with their importance and some examples that lead to heterogeneity creation in structure.

References

1. Hu, M.L., Masoomi, M.Y., Morsali, A., Template strategies with MOFs. *Coord. Chem. Rev.*, *387*, 415–435, 2019.
2. Esrafili, L., Gharib, M., Morsali, A., The targeted design of dual-functional metal–organic frameworks (DF-MOFs) as highly efficient adsorbents for Hg^{2+} ions: synthesis for purpose. *Dalton Trans.*, *48*, 48, 17831–17839, 2019.
3. Miller, T., Wittenberg, J., Wen, H., Connor, S., Cui, Y., Lindenberg, A.M., The mechanism of ultrafast structural switching in superionic copper (I) sulphide nanocrystals. *Nat. Commun.*, *4*, 1369, 2013.
4. Wu, H., Chua, Y.S., Krungleviciute, V., Tyagi, M., Chen, P., Yildirim, T., Zhou, W., Unusual and highly tunable missing-linker defects in zirconium metal–organic framework UiO-66 and their important effects on gas adsorption. *J. Am. Chem. Soc.*, *135*, 28, 10525–10532, 2013.
5. Barin, G., Krungleviciute, V., Gutov, O., Hupp, J.T., Yildirim, T., Farha, O.K., Defect creation by linker fragmentation in metal–organic frameworks and its effects on gas uptake properties. *Inorg. Chem.*, *53*, 13, 6914–6919, 2014.
6. Shearer, G.C., Chavan, S., Ethiraj, J., Vitillo, J.G., Svelle, S., Olsbye, U., Lamberti, C., Bordiga, S., Lillerud, K.P., Tuned to perfection: ironing out the defects in metal–organic framework UiO-66. *Chem. Mater.*, *26*, 14, 4068–4071, 2014.
7. Gharib, M., Esrafili, L., Morsali, A., Retailleau, P., Solvent-assisted ligand exchange (SALE) for the enhancement of epoxide ring-opening reaction catalysis based on three amide-functionalized metal–organic frameworks. *Dalton Trans.*, *48*, 24, 8803–8814, 2019.
8. Vermoortele, F., Vandichel, M., Van de Voorde, B., Ameloot, R., Waroquier, M., Van Speybroeck, V., De Vos, D.E., Electronic effects of linker substitution on Lewis acid catalysis with metal–organic frameworks. *Angew. Chem. Int. Ed.*, *51*, 20, 4887–4890, 2012.

9. St. Petkov, P., Vayssilov, G.N., Liu, J., Shekhah, O., Wang, Y., Wöll, C., Heine, T., Defects in MOFs: A thorough characterization. *ChemPhysChem, 13*, 8, 2025–2029, 2012.
10. Ameloot, R., Vermoortele, F., Hofkens, J., De Schryver, F.C., De Vos, D.E., Roeffaers, M.B., Three-dimensional visualization of defects formed during the synthesis of metal–organic frameworks: a fluorescence microscopy study. *Angew. Chem. Int. Ed., 52*, 1, 401–405, 2013.
11. Cairns, A.B. and Goodwin, A.L., Structural disorder in molecular framework materials. *Chem. Soc. Rev., 42*, 12, 4881–4893, 2013.
12. Cliffe, M.J., Wan, W., Zou, X., Chater, P.A., Kleppe, A.K., Tucker, M.G., Wilhelm, H., Funnell, N.P., Coudert, F.-X., Goodwin, A.L., Correlated defect nanoregions in a metal–organic framework. *Nat. Commun., 5*, 4176, 2014.
13. Sholl, D.S. and Lively, R.P., Defects in metal–organic frameworks: challenge or opportunity? *J. Phys. Chem. Lett., 6*, 17, 3437–3444, 2015.
14. Fang, Z., Bueken, B., De Vos, D.E., Fischer, R.A., Defect-engineered metal–organic frameworks. *Angew. Chem. Int. Ed., 54*, 25, 7234–7254, 2015.
15. Cheetham, A.K., Bennett, T.D., Coudert, F.-X., Goodwin, A.L., Defects and disorder in metal organic frameworks. *Dalton Trans., 45*, 10, 4113–4126, 2016.
16. Shearer, G.C., Chavan, S., Bordiga, S., Svelle, S., Olsbye, U., Lillerud, K.P., Defect engineering: tuning the porosity and composition of the metal–organic framework UiO-66 via modulated synthesis. *Chem. Mater., 28*, 11, 3749–3761, 2016.
17. Bennett, T.D., Cheetham, A.K., Fuchs, A.H., Coudert, F.-X., Interplay between defects, disorder and flexibility in metal-organic frameworks. *Nat. Chem., 9*, 1, 11, 2017.
18. Ren, J., Ledwaba, M., Musyoka, N.M., Langmi, H.W., Mathe, M., Liao, S., Pang, W., Structural defects in metal–organic frameworks (MOFs): Formation, detection and control towards practices of interests. *Coord. Chem. Rev., 349*, 169–197, 2017.
19. Dissegna, S., Epp, K., Heinz, W.R., Kieslich, G., Fischer, R.A., Defective Metal-Organic Frameworks. *Adv. Mater., 30*, 37, 1704501, 2018.
20. Yuan, L., Tian, M., Lan, J., Cao, X., Wang, X., Chai, Z., Gibson, J.K., Shi, W., Defect engineering in metal–organic frameworks: a new strategy to develop applicable actinide sorbents. *Chem. Commun., 54*, 4, 370–373, 2018.
21. Park, H., Kim, S., Jung, B., Park, M.H., Kim, Y., Kim, M., Defect Engineering into Metal–Organic Frameworks for the Rapid and Sequential Installation of Functionalities. *Inorg. Chem., 57*, 3, 1040–1047, 2018.

22. Sarkisov, L., Molecular simulation of low temperature argon adsorption in several models of IRMOF-1 with defects and structural disorder. *Dalton Trans.*, *45*, 10, 4203–4212, 2016.
23. Schrimpf, W., Jiang, J., Ji, Z., Hirschle, P., Lamb, D.C., Yaghi, O.M., Wuttke, S., Chemical diversity in a metal–organic framework revealed by fluorescence lifetime imaging. *Nat. Commun.*, *9*, 1, 1647, 2018.
24. Liu, Y., Klet, R.C., Hupp, J.T., Farha, O., Probing the correlations between the defects in metal–organic frameworks and their catalytic activity by an epoxide ring-opening reaction. *Chem. Commun.*, *52*, 50, 7806–7809, 2016.
25. Llabrés i Xamena, F.X., García Cirujano, F., Corma Canós, A.J.M., Materials, M., An unexpected bifunctional acid base catalysis in IRMOF-3 for Knoevenagel condensation reactions. *Microporous Mesoporous Mater.*, *157*, 112–117, 2012.
26. Øien, S., Wragg, D., Reinsch, H., Svelle, S., Bordiga, S., Lamberti, C., Lillerud, K. P. J. C. G., Design, Detailed structure analysis of atomic positions and defects in zirconium metal–organic frameworks. *Cryst. Growth Des.*, *14*, 11, 5370–5372, 2014.
27. Valenzano, L., Civalleri, B., Chavan, S., Bordiga, S., Nilsen, M.H., Jakobsen, S., Lillerud, K.P., Lamberti, C., Disclosing the complex structure of UiO-66 metal organic framework: a synergic combination of experiment and theory. *Chem. Mater.*, *23*, 7, 1700–1718, 2011.
28. Shöâeè, M., Agger, J.R., Anderson, M.W., Attfield, M.P., Crystal form, defects and growth of the metal organic framework HKUST-1 revealed by atomic force microscopy. *CrystEngComm*, *10*, 6, 646–648, 2008.
29. Carlucci, L., Ciani, G., Moret, M., Proserpio, D.M., Rizzato, S., Polymeric layers catenated by ribbons of rings in a three-dimensional self-assembled architecture: a nanoporous network with spongelike behavior. *Angew. Chem. Int. Ed.*, *39*, 8, 1506–1510, 2000.
30. Shoaee, M., Anderson, M.W., Attfield, M.P., Crystal Growth of the Nanoporous Metal–Organic Framework HKUST-1 Revealed by *In Situ* Atomic Force Microscopy. *Angew. Chem. Int. Ed.*, *47*, 44, 8525–8528, 2008.
31. Taylor, J.M., Komatsu, T., Dekura, S., Otsubo, K., Takata, M., Kitagawa, H., The role of a three dimensionally ordered defect sublattice on the acidity of a sulfonated metal–organic framework. *J. Am. Chem. Soc.*, *137*, 35, 11498–11506, 2015.
32. Ravon, U., Savonnet, M., Aguado, S., Domine, M.E., Janneau, E., Farrusseng, D., Materials, M., Engineering of coordination polymers

for shape selective alkylation of large aromatics and the role of defects. *Micropor. Mesopor. Mater.*, *129*, 3, 319–329, 2010.

33. Cliffe, M.J., Wan, W., Zou, X., Chater, P.A., Kleppe, A.K., Tucker, M.G., Wilhelm, H., Funnell, N.P., Coudert, F.-X., Goodwin, A.L., Correlated defect nanoregions in a metal–organic framework. *Nat. Commun.*, *5*, 4176, 2014.
34. Guillerm, V., Ragon, F., Dan-Hardi, M., Devic, T., Vishnuvarthan, M., Campo, B., Vimont, A., Clet, G., Yang, Q., Maurin, G.A., series of isoreticular, highly stable, porous zirconium oxide based metal–organic frameworks. *Angew. Chem. Int. Ed.*, *51*, 37, 9267–9271, 2012.
35. Tu, B., Pang, Q., Wu, D., Song, Y., Weng, L., Li, Q., Ordered vacancies and their chemistry in metal–organic frameworks. *J. Am. Chem. Soc.*, *136*, 41, 14465–14471, 2014.
36. Sapnik, A., Johnstone, D., Collins, S.M., Divitini, G., Bumstead, A., Ashling, C., Chater, P.A., Johnson, T., Keen, D.A., Bennett, T., Engineering Porosity Through Defects in a Giant Pore Metal–Organic Framework, *ChemRxiv. Prepr.*, 2020.
37. Cliffe, M.J., Hill, J.A., Murray, C.A., Coudert, F.-X., Goodwin, A.L., Defect-dependent colossal negative thermal expansion in UiO-66(Hf) metal–organic framework. *Phys. Chem. Chem. Phys.*, *17*, 11586–11592, 2015.
38. Tealdi, C., Mustarelli, P., Islam, M.S., Layered LaSrGa3O7-Based Oxide-Ion Conductors: Cooperative Transport Mechanisms and Flexible Structures. *Adv. Funct. Mater.*, *20*, 22, 3874–3880, 2010.
39. McCarthy, B.D., Liseev, T., Beiler, A., Materna, K., Ott, S., Facile orientational control of M2L2P SURMOFs on<100> silicon substrates and growth mechanism insights for defective MOFs. *ACS Appl. Mater. Interfaces*, *11*, 41, 38294–38302, 2019.
40. Shekhah, O., Liu, J., Fischer, R., Wöll, C., MOF thin films: existing and future applications. *Chem. Soc. Rev.*, *40*, 2, 1081–1106, 2011.
41. Kind, M. and Wöll, C., Organic surfaces exposed by self-assembled organothiol monolayers: Preparation, characterization, and application. *Prog. Surf. Sci.*, *84*, 7–8, 230–278, 2009.
42. Gutov, O.V., Hevia, M. G. l., Escudero-Adán, E.C., Shafir, A., Metal–organic framework (MOF) defects under control: insights into the missing linker sites and their implication in the reactivity of zirconium-based frameworks. *Inorg. Chem.*, *54*, 17, 8396–8400, 2015.
43. Wu, H., Chua, Y.S., Krungleviciute, V., Tyagi, M., Chen, P., Yildirim, T., Zhou, W., Unusual and highly tunable missing-linker defects in zirconium metal–organic framework UiO-66 and their important effects on gas adsorption. *J. Am. Chem. Soc.*, *135*, 28, 10525–10532, 2013.

44. López-Maya, E., Montoro, C., Rodríguez-Albelo, L.M., Aznar Cervantes, S.D., Lozano-Pérez, A.A., Cenís, J.L., Barea, E., Navarro, J.A., Textile/Metal–Organic-Framework Composites as Self-Detoxifying Filters for Chemical-Warfare Agents. *Angew. Chem. Int. Ed.*, *54*, 23, 6790–6794, 2015.
45. Ardila-Suarez, C., Perez-Beltran, S., Ramirez-Caballero, G., Balbuena, P.B., Technology, Enhanced acidity of defective MOF-808: effects of the activation process and missing linker defects. *Catal. Sci. Technol.*, *8*, 3, 847–857, 2018.
46. Yuan, S., Zou, L., Qin, J.-S., Li, J., Huang, L., Feng, L., Wang, X., Bosch, M., Alsalme, A., Cagin, T., Construction of hierarchically porous metal–organic frameworks through linker labilization. *Nat. Commun.*, *8*, 15356, 2017.
47. Kozachuk, O., Meilikhov, M., Yusenko, K., Schneemann, A., Jee, B., Kuttatheyil, A.V., Bertmer, M., Sternemann, C., Pöppl, A., Fischer, R.A., A Solid-Solution Approach to Mixed-Metal Metal–Organic Frameworks–Detailed Characterization of Local Structures, Defects and Breathing Behaviour of Al/V Frameworks. *Eur. J. Inorg. Chem.*, *2013*, 26, 4546–4557, 2013.
48. Kitagawa, S., Kitaura, R., Noro, S.I., Functional porous coordination polymers. *Angew. Chem. Int. Ed.*, *43*, 18, 2334–2375, 2004.
49. Zhou, H.-C. and Kitagawa, S., Metal-organic frameworks (MOFs). *Chem. Soc. Rev.*, *43*, 5415–5418, 2014.
50. Batten, S.R., Neville, S.M., Turner, D.R., *Coordination polymers: design, analysis and application*, The Royal Society of Chemistry, Thomas Graham House, Science Park, Milton Road, Cambridge, UK, 2008.
51. Bunck, D.N. and Dichtel, W.R., Mixed linker strategies for organic framework functionalization. *Chem.-Eur. J.*, *19*, 3, 818–827, 2013.
52. Chen, X. and Zhang, Q., Recent advances in mesoporous metal-organic frameworks. *Particuology*, *45*, 20–34, 2019.
53. Agirrezabal-Telleria, I., Luz, I., Ortuño, M.A., Oregui-Bengoechea, M., Gandarias, I., López, N., Lail, M.A., Soukri, M., Gas reactions under intrapore condensation regime within tailored metal–organic framework catalysts. *Nat. Commun.*, *10*, 1, 1–8, 2019.
54. Šedivý, L., Čížek, J., Belas, E., Grill, R., Melikhova, O., Positron annihilation spectroscopy of vacancy-related defects in CdTe: Cl and CdZnTe: Ge at different stoichiometry deviations. *Sci. Rep.*, *6*, 20641, 2016.
55. Vermoortele, F., Ameloot, R., Alaerts, L., Matthessen, R., Carlier, B., Fernandez, E.V.R., Gascon, J., Kapteijn, F., De Vos, D.E., Tuning the catalytic performance of metal–organic frameworks in fine chemistry by active site engineering. *J. Mater. Chem.*, *22*, 20, 10313–10321, 2012.

56. Marshall, R.J. and Forgan, R.S., Postsynthetic Modification of Zirconium Metal-Organic Frameworks. *Eur. J. Inorg. Chem., 2016*, 27, 4310–4331, 2016.
57. Karagiaridi, O., Lalonde, M.B., Bury, W., Sarjeant, A.A., Farha, O.K., Hupp, J.T., Opening ZIF-8: a catalytically active zeolitic imidazolate framework of sodalite topology with unsubstituted linkers. *J. Am. Chem. Soc., 134*, 45, 18790–18796, 2012.
58. Karagiaridi, O., Vermeulen, N.A., Klet, R.C., Wang, T.C., Moghadam, P.Z., Al-Juaid, S.S., Stoddart, J.F., Hupp, J.T., Farha, O.K., Functionalized defects through solvent-assisted linker exchange: synthesis, characterization, and partial postsynthesis elaboration of a metal–organic framework containing free carboxylic acid moieties. *Inorg. Chem., 54*, 4, 1785–1790, 2015.
59. Jiang, J.-Q., Yang, C.-X., Yan, X.-P., Postsynthetic ligand exchange for the synthesis of benzotriazole-containing zeolitic imidazolate framework. *Chem. Commun., 51*, 30, 6540–6543, 2015.
60. Szilágyi, P. Á., Serra-Crespo, P., Gascon, J., Geerlings, H., Dam, B., The impact of Post-synthetic linker Functionalization of MOFs on Methane storage: The role of Defects. *Front. Energy Res., 4*, 9, 2016.
61. Deria, P., Mondloch, J.E., Tylianakis, E., Ghosh, P., Bury, W., Snurr, R.Q., Hupp, J.T., Farha, O.K., Perfluoroalkane functionalization of NU-1000 via solvent-assisted ligand incorporation: synthesis and CO2 adsorption studies. *J. Am. Chem. Soc., 135*, 45, 16801–16804, 2013.
62. Bennett, T.D., Todorova, T.K., Baxter, E.F., Reid, D.G., Gervais, C., Bueken, B., Van de Voorde, B., De Vos, D., Keen, D.A., Mellot-Draznieks, C., Connecting defects and amorphization in UiO-66 and MIL-140 metal–organic frameworks: a combined experimental and computational study. *Phys. Chem. Chem. Phys., 18*, 3, 2192–2201, 2016.
63. Nambissan, P.M.G., Characterisation of Ion Implantation-induced Defects in Certain Technologically Important Materials by Positron Annihilation. *Def. Sci. J., 59*, 4, 329–341, 2009.
64. Ameloot, R., Vermoortele, F., Hofkens, J., De Schryver, F.C., De Vos, D.E., Roeffaers, M.B., Three-dimensional visualization of defects formed during the synthesis of metal–organic frameworks: a fluorescence microscopy study. *Angew. Chem., 125*, 1, 419–423, 2013.
65. Walker, A.M. and Slater, B., Comment upon the screw dislocation structure on HKUST-1 {111} surfaces. *CrystEngComm, 10*, 6, 790–791, 2008.

66. Bordiga, S., Regli, L., Bonino, F., Groppo, E., Lamberti, C., Xiao, B., Wheatley, P.S., Morris, R.E., Zecchina, A., Adsorption properties of HKUST-1 toward hydrogen and other small molecules monitored by IR. *Phys. Chem. Chem. Phys.*, *9*, 21, 2676–2685, 2007.
67. Položij, M., Rubeš, M., Čejka, J., Nachtigall, P., Catalysis by Dynamically Formed Defects in a Metal–Organic Framework Structure: Knoevenagel Reaction Catalyzed by Copper Benzene-1, 3, 5-tricarboxylate. *ChemCatChem*, 6, 10, 2821–2824, 2014.
68. Petkov, P.S., Vayssilov, G.N., Liu, J., Shekhah, O., Wang, Y., Wöll, C., Heine, T., Defects in MOFs: a thorough characterization. *ChemPhysChem*, *13*, 8, 2025–2029, 2012.
69. Fang, Z., Dürholt, J.P., Kauer, M., Zhang, W., Lochenie, C., Jee, B., Albada, B., Metzler-Nolte, N., Pöppl, A., Weber, B., Muhler, M., Structural complexity in metal–organic frameworks: Simultaneous modification of open metal sites and hierarchical porosity by systematic doping with defective linkers. *J. Am. Chem. Soc.*, *136*, 27, 9627–9636, 2014.
70. Barin, G., Krungleviciute, V., Gutov, O., Hupp, J.T., Yildirim, T., Farha, O.K., Defect creation by linker fragmentation in metal–organic frameworks and its effects on gas uptake properties. *Inorg. Chem.*, *53*, 13, 6914–6919, 2014.
71. Park, J., Wang, Z.U., Sun, L.B., Chen, Y.P., Zhou, H.C., Introduction of functionalized mesopores to metal–organic frameworks via metal–ligand–fragment coassembly. *J. Am. Chem. Soc.*, *134*, 49, 20110–20116, 2012.
72. Ameloot, R., Vermoortele, F., Hofkens, J., De Schryver, F.C., De Vos, D.E., Roeffaers, M.B., Three-dimensional visualization of defects formed during the synthesis of metal–organic frameworks: a fluorescence microscopy study. *Angew. Chem.*, *125*, 1, 419–423, 2013.
73. Choi, J.S., Son, W.J., Kim, J., Ahn, W.S., Metal–organic framework MOF-5 prepared by microwave heating: Factors to be considered. *Micropor. Mesopor. Mater.*, *116*, 1-3, 727–731, 2008.
74. Tsao, C.S., Yu, M.S., Chung, T.Y., Wu, H.C., Wang, C.Y., Chang, K.S., Chen, H.L., Characterization of Pore Structure in Metal–Organic Framework by Small-Angle X-ray Scattering. *J. Am. Chem. Soc.*, *129*, 51, 15997–16004, 2007.
75. Cliffe, M.J., Wan, W., Zou, X., Chater, P.A., Kleppe, A.K., Tucker, M.G., Wilhelm, H., Funnell, N.P., Coudert, F.X., Goodwin, A.L., Correlated defect nanoregions in a metal–organic framework. *Nat. Commun.*, *5*, 1, 4176, 2014.

76. Yang, S.H., Lin, X., Lewis, W., Suyetin, M., Bichoutskaia, E., Parker, J.E., Tang, C.C., Allan, D.R., Rizkallah, P.J., Hubberstey, P., A partially interpenetrated metal–organic framework for selective hysteretic sorption of carbon dioxide. *Nat. Mater.*, *11*, 8, 710, 2012.
77. Shen, L., Yang, S.W., Xiang, S., Liu, T., Zhao, B., Ng, M.F., Goëttlicher, J., Yi, J., Li, S., Wang, L., Ding, J., Origin of long-range ferromagnetic ordering in metal–organic frameworks with antiferromagnetic dimeric-Cu (II) building units. *J. Am. Chem. Soc.*, *134*, 41, 17286–17290, 2012.
78. Feyand, M., Mugnaioli, E., Vermoortele, F., Bueken, B., Dieterich, J.M., Reimer, T., Kolb, U., De Vos, D., Stock, N., Automated Diffraction Tomography for the Structure Elucidation of Twinned, Sub-micrometer Crystals of a Highly Porous, Catalytically Active Bismuth Metal–Organic Framework. *Angew. Chem.*, *124*, 41, 10519–10522, 2012.
79. Leus, K., Vandichel, M., Liu, Y.Y., Muylaert, I., Musschoot, J., Pyl, S., Vrielinck, H., Callens, F., Marin, G.B., Detavernier, C., Wiper, P.V., The coordinatively saturated vanadium MIL-47 as a low leaching heterogeneous catalyst in the oxidation of cyclohexene. *J. Catal.*, *285*, 1, 196–207, 2012.
80. Nguyen, L.T., Le, K.K., Truong, H.X., Phan, N.T., Metal–organic frameworks for catalysis: the Knoevenagel reaction using zeolite imidazolate framework ZIF-9 as an efficient heterogeneous catalyst. *Catal. Sci. Technol.*, *2*, 3, 521–528, 2012.
81. Furukawa, H., Gandara, F., Zhang, Y.B., Jiang, J., Queen, W.L., Hudson, M.R., Yaghi, O.M., Water adsorption in porous metal–organic frameworks and related materials. *J. Am. Chem. Soc.*, *136*, 11, 4369–4381, 2014.
82. Vo, T.K., Nguyen, V.C., Quang, D.T., Park, B.J., Kim, J., Formation of structural defects within UiO-66 (Zr)-$(OH)_2$ framework for enhanced CO_2 adsorption using a microwave-assisted continuous-flow tubular reactor. *Micropor. Mesopor. Mater.*, *312*, 110746, 2021.

9

Complexity Based on Heterogeneous Pores

Abstract

Most of the reported MOFs are simple homogeneous forms but recently the synthetic MOFs have been made with different multiple components. Using the various building units can generate the heterogeneity in MOFs structure and their pores. It is notable that heterogeneity in pores can cause more structural abilities with the generation of different properties and applications in framework. In this chapter, we present a view of MOFs with multiple kins of pores (macro-, meso-, and micropore) along with synthesis methods such as using mixed linkers. Multiple pore sizes can create more complexity and heterogeneity in frameworks. However, there are multiple kinds of pores from a family with different sizes such as multiple macropores. Such pores in MOFs with heterogeneous structures and their properties can induce high capability in different fields like gas adsorption.

Keywords: Heterogeneous pores, multiple pore sizes, complex MOF structures, selective gas capture

9.1 Heterogeneous Pores

With manipulation of porous materials, the range of pores can change from micro to macro size. These materials can have the combination of different pores as micro-mesopores, meso-macropores, micro-macropores, micro-meso-macroporous supports, and multiple macropores with various morphologies by using the varied synthesis strategies as surfactant/macroporous polymer/colloidal crystal/emulsion templating, breath figures, bioinspiring process,

Ali Morsali and Kayhaneh Berijani. Metal-Organic Frameworks with Heterogeneous Structures, (171–184) © 2021 Scrivener Publishing LLC

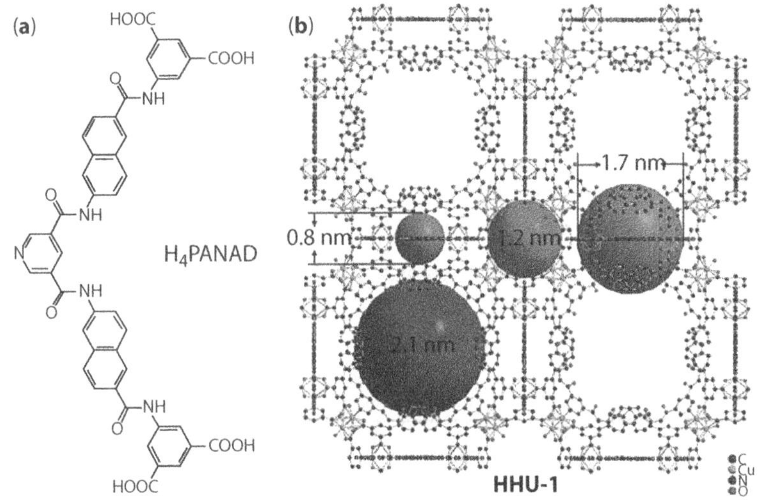

Figure 9.1 (a) H_4PANAD structure. (b) HHU-1 network with varying size pores (two brown atoms illustrate one ellipsoid-like pore). For clarity, H atoms have not been shown [42].

and phase separation [1–9]. MOFs as porous materials with various organic and inorganic components can have diverse porosities with different pore sizes (Figure 9.1). The usual strategy to synthesize of MOFs with different pores is template effect (hard or soft template) that with removing it, ordered or irregular pores can be generated.

Some of the properties in MOFs can induce complexity to framework due to the variety in the structural components. Porous MOFs are usually homogeneous structures with one type of pore that is repeated in structure without any complexity but there are methods to make complex architectures. One of the synthetic methods that can generate heterogeneity in crystals of MOFs is the creation of the different pores in plannings. With this phrase "heterogeneous pores in MOFs", the first thing that comes into our minds is that we have one MOF which has two different kinds of pores. MOFs are fascinating due to their structural properties especially with having various pores [10, 11].

This level from complexity can add new properties such as storage [12]. The stable MOFs can also have the heterogeneous pores without losing their pristine crystalline nature. The heterogeneously structure of pores affects MOFs' applications such as gas storage, drug delivery, and insertion of external particles. These kinds of MOFs with the two or three different types of porosities (micro, meso, and macro) that are juxtaposed represent unique porous MOFs as significant synthetic compounds. Even, based on the designing in the organic section (organic linker), cavities can be formed from one kind but with more than one kind of different sizes that can induce the heterogeneity to cavities of framework [13, 14]. Also, an organic linker can have the different conformational isomers that can combine with the various SBUs to construct MOFs with diverse pores [15]. This simple and interesting method employed to construct a MOF with three different SBUs and four different cages by using pyrazolecarboxylate linker and two distinct metal ions (Cu and Zn) [16]. The heterogeneous pores can be synthesized with hierarchical structures, too. One example of these MOFs is PCN-80 as a Cu-MOF with rht-topology that has been synthesized through the solvothermal reaction [13]. The used organic ligand in this structure is 9,9',9'',9'''-([1,1'-biphenyl]-3,3',5,5'-tetrayl)tetrakis(9H-carbazole-3,6-dicarboxylate) with two adjacent carboxylates groups that the bridging angle of both of them is 90°. In this MOF, the kind of cavities is microporous but their size is diverse (16.6, 13.4, and 7.6 Å). In these kinds of syntheses, BET surface area of MOFs can increase and their cages nature can make MOFs as materials with property of gas storage. The different size, shape, and nature of designed cavities can impress on the influence and affinity of specific molecules into framework [17–20]. Inducing heterogeneity to pores structures, creating of heterogeneously pores, is seen to different forms with different methods. Herein, we expand this subject with more instances and the other conditions that lead to complexity in MOFs that some of them are implied in the following.

The yolk-shell synthesis can lead to the production of the void space even with the larger void space by using the spray drying method inside shell [21, 22]. MOFs can be synthesized with mesopore core and micropore shell [23]. As well, with adding the beads of polystyrene into MOFs meso-/macroscopic structures and 3D macroporous Al_2O_3, nano-sized MOF crystals of [Al(OH)(NDC)] were synthesized with different pores [24, 25]. One instance with further explanation: in 2018, Chen *et al.* synthesized 3D single crystals MOF with ordered micro/macropores by using the specified amount of imidazole linker/zinc metal, polystyrene, and MeOH/NH_3·H2O [26]. This MOF was formed during three steps: (a) the impregnation of ZIF-8 components, (b) crystallization of the first stage product in the presence of the mixture of wood alcohol (MeOH) and ammonia, and (c) in the last step, the polystyrene removed as template and final product, SOM-ZIF-8, was created. XRD shows the preserve of the crystalline structure with comparing to parent framework. As mentioned in the above, MOFs exist with two types of pores from one kind pore such as meso or micro that their difference is in the internal diameters of cages or dimensions like the synthesis of meso-MOFs MIL-100 and MIL-101 with the same topology and different mesoporous cages. Another approach to prepare MOFs with various cavities is, for example, the cage-like micropores in frameworks of MOFs that have various shapes and dimensions [27].

A microporous In(III)-MOF was reported with different pores that had various dimensions. It is interesting that the heterogeneous pores can be observed due to the defect in the framework nature with missing linker strategy. Through this method, pores are generated with diverse sizes. With the further studies, it has been observed that "missing node" can be produced as defect in MOF which induces new capabilities to framework (Figure 9.2) [28, 29]. Hydrolytic process is a new strategy to prepare of hierarchical heteropores MOFs that during this process, two main factors temperature and time can control the pores size. This kind of MOFs can be used to stabilize different chemical compounds as macromolecular biological molecules and they can be employed in metabolic processes [29].

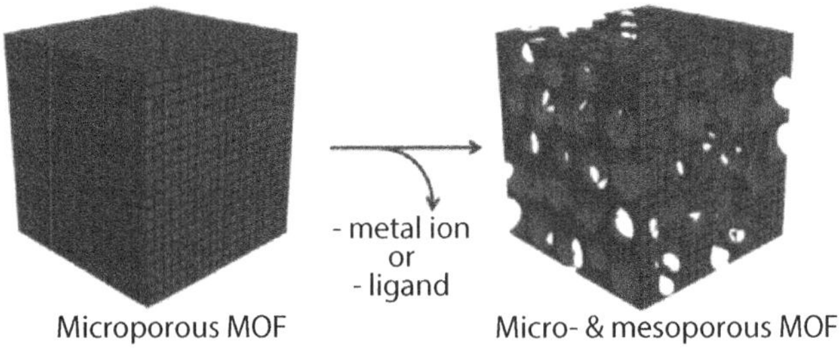

Figure 9.2 Heterogeneous pores due to missing node or ligand [28, 29].

One of the applications that are investigated on this kind of MOF is gas adsorption ability in details such as different temperatures [30]. SEM images are evidences for the presence of the arrangement of pores in different direction. BET analysis confirms the existence of pores with the different sizes. The other characterization technique, high-angle annular dark-field scanning transmission electron microscopy, investigates the prepared pores in details. The interesting point is that the catalytic activity in these MOFs was studied, too. The excellent obtained results show that these valuable researches are worthy publishing in Science journal. As mentioned above, another case of MOFs with different pores architecture can be micro-, meso- and macro-size. MOFs due to tunablity in their structure, can be synthesized with these kinds of pores and promising features. This kind of the various pore-size distribution in MOFs is very significant because they can induce advantages and abilities to framework. One of the most important advantages is easy mass transfer specially in the catalytic processes. There are several synthesis methods with variations in preparing conditions to create these kinds of MOFs. **Solvothermal route** is one of them that was used by Zhang *et al.* for synthesis of a multiporous MOF that is defective [31]. Its inorganic and organic parts were Zn and bpydc. CO_2-expanded DMF was also utilized. After experimental details for the synthesis that authors reported, they characterized the new obtained structure by different techniques such as SEM, TEM, N2 adsorption-desorption isotherm, XPS, XANES and EXAFS. In the first step, micro-, meso- and macropores were identified and their size were measured (micro, meso and macro diameters: 0.73, 3.9 and 100 nm, respectively). In agreement with the obtained finding about micropore size by N2 adsorption/desorption, single-crystal diffraction data showed similar amount. Two processes of particles assembly and cross-linking affect on the production of large pores. This as-synthesized MOF due to its composition could act as an oxidative photocatalyst of amines under ambient and mild catalytic conditions with high selectivity. Surely, the existence of heterogeneous pores and heterogeneity within structure has positive effect on the considered MOF specially by producing improved surface area, facile entering reactant molecules with large size, access to channels of

Zn-bpydc in catalytic process. Indeed, herein, exposing catalytic centers and diffusion or transport of substrates or products in diffusion-influenced catalytic reactions are owing to hierarchical porous. ***Supercapacitors*** are also very valuable in the energy field because they can save energy. In these processes such as electrochemical and capacitive storage, porous materials with heterogeneous pores in their structure are important. So, the design of systems with this capability, is a key and an important factor. Storage mechanisms of supercapacitors are two kinds: (i) double layer capacitance and (ii) pseudo-capacitance. In the first case, porous materials and in another state, transition metal oxides are usually investigated [32]. Given, transition metal oxides have limitations in reversibility and structural stability, therefore supercapacitors with the first case are investigated because they have long lifetime and suitable energy density. Recently, MOFs and MOF-derived porous materials have received considerable attention in

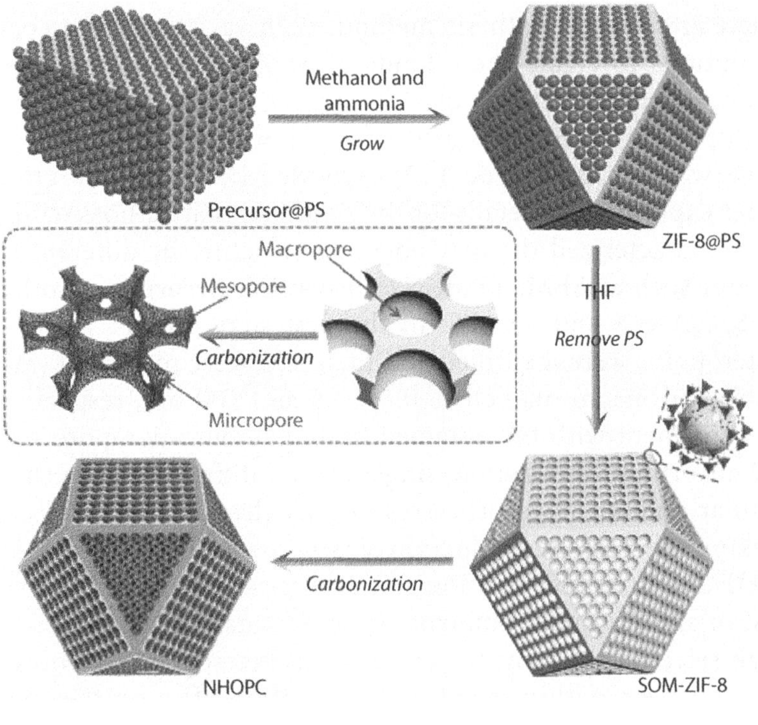

Figure 9.3 An example of MOF-based porous carbon (NHOPC) with multiporosity synthesis as a supercapacitor [33].

this area. N-doped macro-meso-micro carbon materials are one of them that in this field can be utilized. Newly, Qiao, Mamat and co-workers have reported NHOPC that has been derived from macroporous zeolitic MOF (ZIF-8) (Figure 9.3) [33].

Large SSA and the existence of heterogeneous pores are structural features that helped to convert this compound to a supercapacitor with enhanced energy density, high capacitance and power density. Good synergic effect of these properties showed the improved electron and mass transfer pathway as well as positive effect of N-doping to the structure. So, the absence of materials with excellent engineering design and properties generate technological problems, that we somewhat know their known difficulties and problems which limit performance of materials. For example, it has been reported that porous materials with small pores such as micropore can have negative effect on the vital factors of diffusion and capacitance, while mesopores can have positive effect on this process [34]. Or macropores that are beneficial for ion-based electrolytes storage and micropores in structure that can produce large specific surface area which has effect on supercapacitors ability. So, certainly, the multiple pores as heterogeneoty agents in structure are showing great promise for developing new types of porous materials with desirable performance. In contrast to the kinds of heterogeneous porosities that we discussed in the results above, there are several other kinds of heteropore MOFs that do not belong to the above categories and we want to introduce them. As above explained, there are pores with different scales: micro smaller than 2 nm, meso with useable range: 2–50 nm and macro greater than 50 nm. Whereas these pores can be in the specified range with various sizes (like small meso-large meso), so the existence of these kinds of pores, from one category with different sizes within the structure, can induce heterogeneity into structure. It should be said that these kinds of pores are also observed in hierarchical non-MOF porous materials [35–37]. Prior to using MOFs, their structural features like porosity should be investigated as that it is a key and logic property. The production of heterogeneous pores with tunable sizes from one class is significant and can provide useful and attractive heterogeneous porous platforms with the considerable heterogeneity. For example, we start with mesoMOF

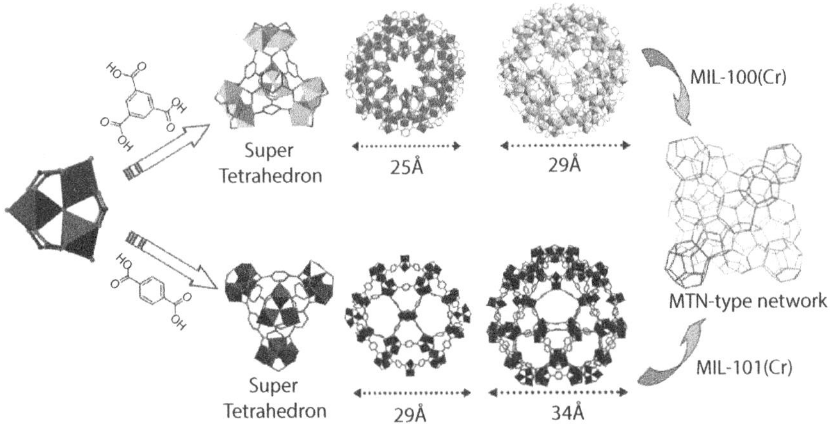

Figure 9.4 Examples of Cr-MILs as mesoMOFs with two types of mesocages [43, 44].

materials that pores structure and their shape can appear in two general kinds: (i) cage- and or (ii) channel-kind. Chromium and terephthalate-based MOFs, MIL-101s, are better example of cage-kind to introduce of this type of materials (Figure 9.4).

It is interesting that heterogeneous pores in covalent organic framework (COF) are also observed with diverse sizes as micro/meso size [38]. These kinds of materials are very important due to their applications, for instance, gas adsorption and heterogeneous catalysts [39, 40]. Even, metal-organic materials (MOMs) that are derived from MOFs and MOPs (metal-organic polyhedra) can be synthesized with three different pores [41]. Truly, these innovations are admirable and respectful.

9.2 Conclusion

To now, many metal-organic frameworks as porous materials have been synthesized by using the inorganic and organic components. Most of the reported structures are simple homogeneous frameworks but recently the synthetic MOFs have been made with multiple different components. Using the various building units can generate the heterogeneity in MOF pores. It is notable that heterogeneity in pores can cause more structural abilities along with the generation of different properties. These multiple different pores can induce properties and applications into MOFs framework through the produced heterogeneity. For example, micro- and mesopores, micro- and macropores, meso- and macropores, even three kinds of different pores can exist in frameworks, simultaneously. The presence of heterogeneous pores in MOFs due to complexity can be an important avenue in MOFs field.

References

1. Razavi, S.A.A., Masoomi, M.Y., Islamoglu, T., Morsali, A., Xu, Y., Hupp, J.T., Farha, O.K., Wang, J., Junk, P.C., Improvement of Methane–Framework Interaction by Controlling Pore Size and Functionality of Pillared MOFs. *Inorg. Chem.*, *56*, 5, 2581–2588, 2017.
2. Fratzl, P., Biomimetic materials research: what can we really learn from nature's structural materials? *J. R. Soc. Interface*, *4*, 15, 637642, 2007.
3. Fratzl, P. and Weinkamer, R., Nature's hierarchical materials. *Prog. Mater. Sci.*, *52*, 8, 1263–1334, 2007.
4. Lakes, R., Materials with structural hierarchy. *Nature*, *361*, 6412, 511, 1993.
5. Hashemi, L. and Morsali, A., A new lead (II) nanoporous three-dimensional coordination polymer: pore size effect on iodine adsorption affinity. *CrystEngComm*, *16*, 23, 4955–4958, 2014.
6. Messersmith, P.B., Multitasking in tissues and materials. *Science*, *319*, 5871, 1767–1768, 2008.
7. Ghasempour, H., Tehrani, A.A., Morsali, A., Wang, J., Junk, P.C., A novel 3D pillar-layered metal-organic framework: Pore-size-dependent catalytic activity and CO_2/N_2 affinity. *Polyhedron*, *180*, 114422–114429, 2020.
8. Razavi, S.A.A. and Morsali, A., Ultrasonic-Assisted Linker Exchange (USALE): A Novel Post-Synthesis Method for Controlling the Functionality, Porosity, and Morphology of MOFs. *Chem.-Eur. J.*, *25*, 46, 10876–10885, 2019.

9. Su, B.-L., Sanchez, C., Yang, X.-Y., *Hierarchically structured porous materials: from nanoscience to catalysis, separation, optics, energy, and life science*, John Wiley & Sons, Weinheim, Germany, 2012.
10. Long, J.R. and Yaghi, O.M. (Eds.), Themed issue: Metal-Organic Frameworks. *Chem. Soc. Rev.*, *38*, 5, 1201–1508, 2009.
11. Férey, G. and Serre, C., Large breathing effects in three-dimensional porous hybrid matter: facts, analyses, rules and consequences. *Chem. Soc. Rev.*, *38*, 5, 1380–1399, 2009.
12. Sudik, A.C., Côté, A.P., Wong-Foy, A.G., O'Keeffe, M., Yaghi, O.M., A metal–organic framework with a hierarchical system of pores and tetrahedral building blocks. *Angew. Chem. Int. Ed.*, *45*, 16, 2528–2533, 2006.
13. Lu, W., Yuan, D., Makal, T.A., Li, J.R., Zhou, H.C.A., Highly Porous and Robust (3, 3, 4)-Connected Metal–Organic Framework Assembled with a 90° Bridging-Angle Embedded Octacarboxylate Ligand. *Angew. Chem. Int. Ed.*, *51*, 7, 1580–1584, 2012.
14. Lu, W., Yuan, D., Makal, T.A., Wei, Z., Li, J.-R., Zhou, H.-C., Highly porous metal–organic framework sustained with 12-connected nanoscopic octahedra. *Dalton Trans.*, *42*, 5, 1708–1714, 2013.
15. Li, J.-R., Yu, J., Lu, W., Sun, L.-B., Sculley, J., Balbuena, P.B., Zhou, H.-C., Porous materials with pre-designed single-molecule traps for CO 2 selective adsorption. *Nat. Commun.*, *4*, 1538, 2013.
16. Tu, B., Pang, Q., Ning, E., Yan, W., Qi, Y., Wu, D., Li, Q., Heterogeneity within a mesoporous metal–organic framework with three distinct metal-containing building units. *J. Am. Chem. Soc.*, *137*, 42, 13456–13459, 2015.
17. Zhang, M., Bosch, M., Gentle III, T., Zhou, H.-C., Rational design of metal–organic frameworks with anticipated porosities and functionalities. *CrystEngComm*, *16*, 20, 4069–4083, 2014.
18. Li, J.-R., Kuppler, R.J., Zhou, H.-C., Selective gas adsorption and separation in metal–organic frameworks. *Chem. Soc. Rev.*, *38*, 5, 1477–1504, 2009.
19. Li, J.-R., Sculley, J., Zhou, H.-C., Metal–organic frameworks for separations. *Chem. Rev.*, *112*, 2, 869–932, 2011.
20. Li, J.-R., Ma, Y., McCarthy, M.C., Sculley, J., Yu, J., Jeong, H.-K., Balbuena, P.B., Zhou, H.-C., Carbon dioxide capture-related gas adsorption and separation in metal-organic frameworks. *Coord. Chem. Rev.*, *255*, 15–16, 1791–1823, 2011.

21. Kuo, C.-H., Tang, Y., Chou, L.-Y., Sneed, B.T., Brodsky, C.N., Zhao, Z., Tsung, C.-K., Yolk–shell nanocrystal@ZIF-8 nanostructures for gas-phase heterogeneous catalysis with selectivity control. *J. Am. Chem. Soc.*, *134*, 35, 14345–14348, 2012.
22. Carné-Sánchez, A., Imaz, I., Cano-Sarabia, M., Maspoch, D., A spray-drying strategy for synthesis of nanoscale metal–organic frameworks and their assembly into hollow superstructures. *Nat. Chem.*, *5*, 3, 203, 2013.
23. Choi, K.M., Jeon, H.J., Kang, J.K., Yaghi, O.M., Heterogeneity within order in crystals of a porous metal–organic framework. *J. Am. Chem. Soc.*, *133*, 31, 11920–11923, 2011.
24. Reboul, J., Furukawa, S., Horike, N., Tsotsalas, M., Hirai, K., Uehara, H., Kondo, M., Louvain, N., Sakata, O., Kitagawa, S., Mesoscopic architectures of porous coordination polymers fabricated by pseudomorphic replication. *Nat. Mater.*, *11*, 8, 717, 2012.
25. Comotti, A., Bracco, S., Sozzani, P., Horike, S., Matsuda, R., Chen, J., Takata, M., Kubota, Y., Kitagawa, S., Nanochannels of two distinct cross-sections in a porous Al-based coordination polymer. *J. Am. Chem. Soc.*, *130*, 41, 13664–13672, 2008.
26. Shen, K., Zhang, L., Chen, X., Liu, L., Zhang, D., Han, Y., Chen, J., Long, J., Luque, R., Li, Y., Ordered macro-microporous metal-organic framework single crystals. *Science*, *359*, 6372, 206–210, 2018.
27. Gu, J.M., Hong, J.Y., Won, Y.S., Park, S.S., Huh, S., Experimental and Theoretical Investigations of CO_2 Sorption by a 3D In-MOF with Multiple 1D Channels. *Eur. J. Inorg. Chem.*, *2015*, 24, 4038–4043, 2015.
28. Barin, G., Krungleviciute, V., Gutov, O., Hupp, J.T., Yildirim, T., Farha, O.K., Defect creation by linker fragmentation in metal–organic frameworks and its effects on gas uptake properties. *Inorg. Chem.*, *53*, 13, 6914–6919, 2014.
29. Kim, Y., Yang, T., Yun, G., Ghasemian, M.B., Koo, J., Lee, E., Cho, S.J., Kim, K., Hydrolytic transformation of microporous metal–organic frameworks to hierarchical micro-and mesoporous MOFs. *Angew. Chem. Int. Ed.*, *54*, 45, 13273–13278, 2015.
30. Huh, S., Kwon, T.-H., Park, N., Kim, S.-J., Kim, Y., Nanoporous In-MOF with multiple one-dimensional pores. *Chem. Commun.*, 33, 4953–4955, 2009.

31. Sha, Y., Zhang, J., Tan, D., Zhang, F., Cheng, X., Tan, X., Zhang, B., Han, B., Zheng, L., Zhang, J., Hierarchically macro–meso–microporous metal–organic framework for photocatalytic oxidation. *Chem Commun., 56,* 73, 10754–10757, 2020.
32. Zhong, S., Zhan, C., Cao, D., Zeolitic imidazolate framework-derived nitrogen-doped porous carbons as high performance supercapacitor electrode materials. *Carbon, 85,* 51–59, 2015.
33. Wang, Y., Qiao, M., Mamat, X., Nitrogen-doped macro-meso-micro hierarchical ordered porous carbon derived from ZIF-8 for boosting supercapacitor performance. *Appl. Surf. Sci., 540,* 148352, 2021.
34. M. Jiang, X. Cao, D. Zhu, Y. Duan, J. Zhang, Hierarchically porous N-doped carbon derived from ZIF-8 nanocomposites for electrochemical applications. *Electrochimica Acta, 196,* 699–707, 2016.
35. Zhao, T., Li, S., Xiao, Y.X., Janiak, C., Chang, G., Tian, G., Yang, X.Y., Template-free synthesis to micro-meso-macroporous hierarchy in nanostructured MIL-101 (Cr) with enhanced catalytic activity. *Science China Materials, 64,* 1, 252–258, 2021.
36. Fonseca, J., Choi, S., Synthesis of a novel amorphous metal organic framework with hierarchical porosity for adsorptive gas separation. *Micropor. Mesopor. Mater., 310,* 110600, 2021.
37. Wang, Q., Zhangsun, H., Zhao, Y., Zhuang, Y, Xu, Z., Bu, T., Li, R., Wang, L., Macro-meso-microporous carbon composite derived from hydrophilic metal-organic framework as high-performance electrochemical sensor for neonicotinoid determination. *J. Hazard. Mater., 411,* 125122, 2021.
38. Zhou, T.-Y., Xu, S.-Q., Wen, Q., Pang, Z.-F., Zhao, X., One-step construction of two different kinds of pores in a 2D covalent organic framework. *J. Am. Chem. Soc., 136,* 45, 15885–15888, 2014.
39. Neti, V.S.P.K., Wu, X., Deng, S., Echegoyen, L., Selective CO_2 capture in an imine linked porphyrin porous polymer. *Polym. Chem., 4,* 17, 4566–4569, 2013.
40. Fang, Q., Gu, S., Zheng, J., Zhuang, Z., Qiu, S., Yan, Y., 3D microporous base-functionalized covalent organic frameworks for size-selective catalysis. *Angew. Chem. Int. Ed., 53,* 11, 2878–2882, 2014.
41. Lee, J., Kwak, J.H., Choe, W., Evolution of form in metal–organic frameworks. *Nat. Commun., 8,* 14070, 2017.
42. Lu, Z., Zhang, J., He, H., Du, L. and Hang, C., A mesoporous (3, 36)-connected txt-type metal–organic framework constructed by using a naphthyl-embedded ligand exhibiting high CO 2 storage and selectivity. *Inorg. Chem. Front., 4,* 4, 736-740, 2017.

43. Férey, G., Mellot-Draznieks, C., Serre, C., Millange, F., Dutour, J., Surblé, S. and Margiolaki, I., A chromium terephthalate-based solid with unusually large pore volumes and surface area. *Science, 309,* 5743, 2040-2042, 2005.
44. Férey, G., Serre, C., Mellot-Draznieks, C., Millange, F., Surblé, S., Dutour, J. and Margiolaki, I., A hybrid solid with giant pores prepared by a combination of targeted chemistry, simulation, and powder diffraction. *Angew. Chem. Int. Ed., 116,* 46, 6456-6461, 2004.

10
Complexity Based on Mixed MOFs

Abstract
One of the methods that can produce complexity in MOFs is mixed MOF formation. They can be considered as complex structures that deviation from uniformity in these materials can be observed. Mixed MOFs as MOF-on-MOF heterostructures, core-shell MOF structures, and multi-component SURMOFs (surface-mounted MOFs) are complex MOFs that have heterogeneity in their structures. So, in this chapter, we explain about them, their design, properties, and applications along with related examples.

Keywords: Core-shell MOF, SURMOFs, mixed MOFs, complex frameworks, heterostructures

10.1 Complex Mixed MOFs

The complexity in chemical structures and constructing complex compounds without losing properties with the control of size and morphologies are very important, difficult, and challenging. Given that we are talking about the complexity in MOFs, we know that not only MOFs have complexity but also the other compounds can be complex such as nanoparticles. In 1962, Aleksander Abramovich Grinberg published a book about the chemistry of complex compounds. Recently, a research team of chemists with Schaak as leader at Penn State has succeeded to present a design for creating of complexity in nanoparticles with easy process. The synthesis of these compounds in the laboratory is a difficult way but they propose a simple route to generate complex nanoparticles that are inaccessible. The result of their investigations was published in the journal *Science* that is very valuable [1]. Day by day, the rate of progress

Ali Morsali and Kayhaneh Berijani. Metal-Organic Frameworks with Heterogeneous Structures, (185–198) © 2021 Scrivener Publishing LLC

of the research in this area is increasing. The conversion of mixed MOFs to MOF composite/hybrid form has been attracted as a significant research with high performance and various applications [2–5].

Of course, there are not many reports about mixed MOFs, despite they are making big jump in the useful applications [6]. Because this topic has extra challenges, it can be solved by great and continuous efforts. Here, we imply to more related examples. There are different forms of core-shell heterostructures and MOF-on-MOF hybrids as MOF thin films [7]. We found out that in addition to the manipulation of MOFs backbone that can create complexity in MOFs structures, the mixing of MOFs with core-shell method creates structural complexity [8–10]. There are several methods to construct these kinds of MOFs hybrids that often are stepwise methods and had some limitations so they could be unsuitable. Epitaxial growth as a one-pot synthesis has been introduced to prepare MOF-on-MOF hybrids. Liquid-phase epitaxial growth can be employed for construction of heterostructure MOFs on the functionalized surface ($[Zn_4O(carboxypyrazolate)_3]_n$-based MOF thin films (Zn-MI and Zn-DM) with a SAM-functionalized Au-coated substrate) [11]. In design of this kind of MOF composites, there is tunability in both of core and shell, too. The nucleation kinetic analysis helps to construct this kind of composites. The encapsulation process during the synthesis of MOF composites can lead to complexity and even it can increase complexity [12–17]. With the further investigations, it has been specified that core-shell MOFs can enhance complexity due to the existence functions that can be in core MOF or shell MOF. The kind of core MOF or shell MOF can be the same or different such as bio-MOF composite bio-MOF-11@bio-MOF-14 (Figure 10.1) [18–24].

The other kind of core-shell composites has the same core and shell such as MIL-68@MIL-68-Br or MIL-68@MIL-68-NDC but their difference is based on functional groups [18]. In some of the core-shell MOF composites, the porosity of core and shell are different like mesoporous cores and microporous shells. Sometimes, the synthesis of core-shell MOFs hybrid can be formed by using MOFs that have two distinct metals. Whole of the mentioned points and their effects were investigated in the other sections that they can

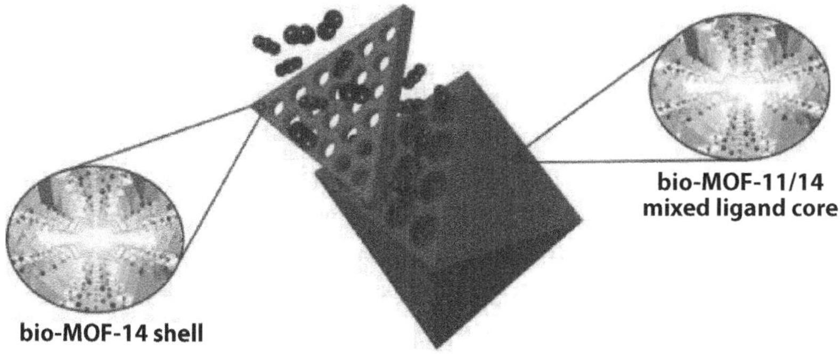

Figure 10.1 Schematic representation of bio-MOF composite [21].

generate complexity in MOFs, from functionalities to mixed metals. Certainly, the kind of functional group affects the growth direction and we know that the varied functional groups can induce heterogeneity into structure of MOFs. Mixed ZIFs composites as core-shell ZIFs have been also synthesized and they can show valuable application and properties rather than the single counterparts [25]. In most of the investigations, mixed MOFs can be produced *via* epitaxial growth but heteroepitaxial growth strategy is also used to fabricate of core-shell MOFs.

The heteroepitaxial strategy can be used to grow isoreticular MOFs on the other isoreticular MOFs to single crystal form. Amine compound in this strategy plays main role because it helps to grow shell MOF that single crystal XRD confirmed heteoepitaxially growing shell on the core [7]. The growth of a MOF on the other MOF can be investigated by synchrotron X-ray analysis and surface plasmon resonance spectroscopy. As well, the results show that the various thin films can be prepared with this method and they improved features and practical applications. For example, Jeong *et al.* reported isoreticular metal-organic framework (IRMOF) synthesis by heteroepitaxial growth that is growing one IRMOF on another at the single crystal level. In this synthesis, the presence of an amine compound has essential role in the heteroepitaxial growth of IRMOFs by preventing IRMOF core from dissolving during epitaxial growth step. Single-crystal X-ray diffraction proved the performing this process: epitaxy of shell crystals grown on core crystals surface. With this method, IRMOF-3 crystals could grow on the surface of IRMOF-1 seed crystal layers. These results show that this preparing method can be used to synthesize novel hybrids along with their thin films with advanced performances (Figure 10.2). To investigate hybrid IRMOF-3@1 and IRMOF-1@3 crystals, their optical micrographs were studied (Figure 10.3). IRMOF-3 growth on IRMOF-1 crystals surface was recognized owing to their different colors. Liu and Wöll published a review about the surface-supported metal-organic framework thin films in 2017 [26]. The progress about MOF thin films and the strategies of their preparing have been widely explained [27]. In most reports, the immobilization of one kind of MOF on surface has been investigated but the growth or deposition of two different types of MOFs onto two- or three-dimensional surfaces is a method for the producing of MOF composites along with complexity that has been less spoken about. In 2017, Jansen, Budd, and co-workers investigated the different layers growth from two distinct types of MOFs, ZIF-8, and HKUST-1 onto a polymer membrane substrate. The obtained observations show that the structure of synthesized composite can maintain high permeability [28]. These hybrid materials can play

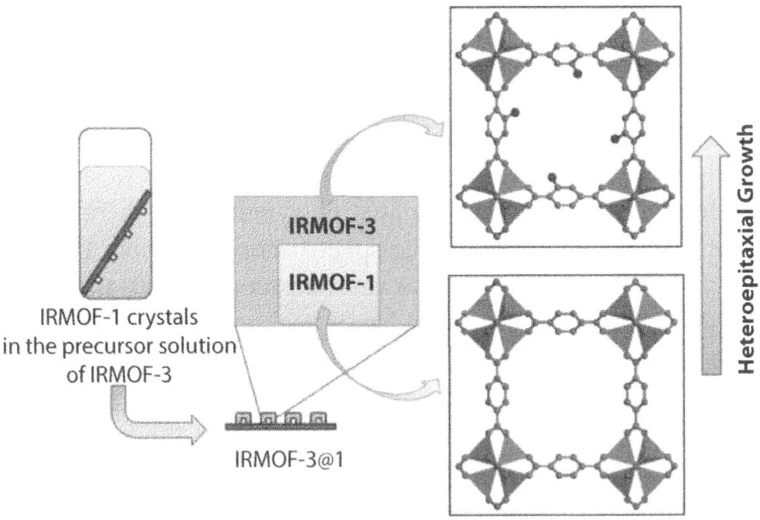

Figure 10.2 Heteroepitaxially grown hybrid IRMOF-3@1 and structures of IRMOF-1 and IRMOF-3 (zinc: purple; nitrogen: blue; oxygen: red; carbon: gray. Hydrogen atoms have not been shown) [7].

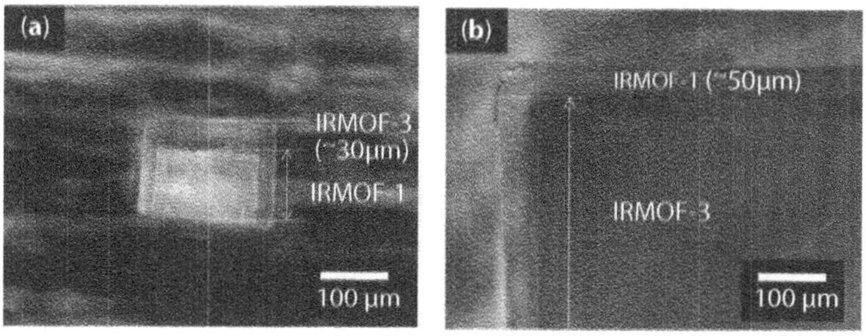

Figure 10.3 Optical micrographs: (a) hybrid IRMOF-3@1, (b) IRMOF-1@3. (IRMOF-1: transparent; IRMOF-3: brownish) [7].

membrane role especially in the industrial processes [29–33]. The used method to synthesize of MOF thin films (MOF-on-MOF) is liquid-phase epitaxy (LPE). This strategy helps to construct the thin films with homogeneous structure. In this method, the used surface should have the ability of functionalization with the different functional groups as terminated OH- and -COOH. The growth of different MOFs can be done on the designed SAMs as a powerful method for different potential chemical applications [34, 35].

The immobilization of thin films of ZIFs has happened, too. The thickness of thin films can be controlled due to the growth time number and rates [36]. The fabrication of hetero-systems and distinct MOF thin films through heteroepitaxial growth is possible that can induce heterogeneity and complexity into MOFs [19, 20, 37]. It is interesting to know that chiral SURMOFs were investigated, too. For the first time, Liu *et al.* synthesized $[Zn_2(Dcam)_2(dabco)]$ that it was used for uptake of enantiomeric molecules [38]. Even, with the existence of all complexities, hetero-layered SURMOF $[Cu_2(Lcam)_2(dabco)]$(L-MOF) on $[Cu_2(Dcam)_2(dabco)]$ (DMOF) was synthesized with Cu^{2+} dimmers, chiral linkers, and dabco as a pillar linker [38, 39]. So, these researches are valuable and significant. Recently, Xiao *et al.* reported bimetal mixed MOFs [based on Co/Zn (*CZ-BDO*), Co/Ni (*CN-BDO*), and Ni/Zn (*NZ-BDO*)] through solvothermal method. After synthesis, the amount of metals of mixed MOFs was investigated by using ICP-OES. Also, their PXRD patterns were also in line with those of monometallic samples. With XPS analysis, chemical states on the samples

Table 10.1 Carbon dioxide cycloaddition with various epoxides by using CZ-BDO.[a]

Entry	Epoxides	Products	Co-catalyst	Temperature	Conversion	Yield
1[a]	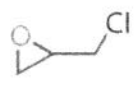	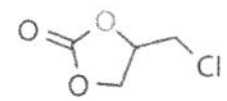	–	100	99.31	97.05
2[a]		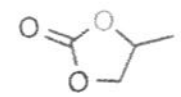	–	100	98.13	90.97
3[a]	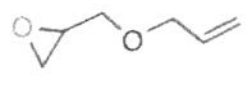	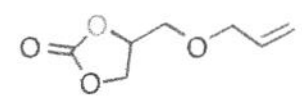	–	100	42.07	40.16
4[a]	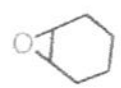	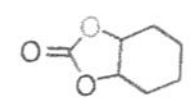	–	100	16.89	16.26
5[b]		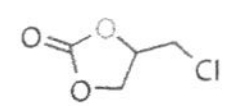	2,6-Lutidine	110	98.0	–
6[c]		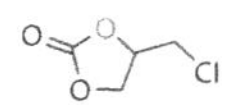	*N*Bu_4NBr	70	–	99.0
7[d]		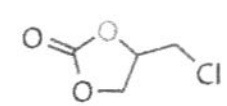	TBAB	100	–	95
8[e]	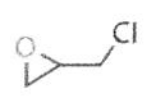	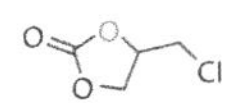	–	130	–	81–86

[a]CO_2 coupling with various epoxides: epoxides 20.0 g, 100°C, 3.0 MPa, 5 h, catalyst at 0.5 wt% of epoxides. [b]ECH 0.98 MPa, 18 h, 0.015 mol% ECH catalyst, 2 mol% ECH co-catalyst. [c]ECH 20 mmol, 1.0 MPa, 12 h, 2.0 mol% ECH catalyst, 2.5 mol% ECH co-catalyst. [d]ECH, 0.1 MPa, 18 h, 0.2 mol% ECH catalyst, 0.4 mol% TBAB. [e]ECH 0.143 mol, 1.0 MPa, 5 h, 0.477 mmol catalyst.

surface were investigated. Techniques SEM, EDS, Raman, N2 adsorption, XRD, and TG-DTG were also employed. According to the obtained findings and nature of these materials, chemical fixation of CO_2 into cyclic carbonates as an important catalytic reaction was investigated. The catalytic results showed that CZ-BDO mixed MOF was superior to the other MOFs. The obtained high yield can be related to the synergistic effect between Co and Zn in catalysis. This suggestion was confirmed by evaluation of NH_3-TPD [40]. Results of carbon dioxide cycloaddition with various epoxides by using CZ-BDO have been shown in Table 10.1 along with catalytic plausible mechanism in Scheme 10.1. Another application of this kind of material is oxygen evolution reaction (OER) that is very important in hydrogen production [41–50]. S-doping NiCo-MOF (NiCoS-MOF) was successfully synthesized by *in situ* method by solvothermal process. According to chemical structure of NiCoS-MOF, its electrocatalytic performance as OER electrodes was investigated. High stability of structure without any significant collapse after 1,000 cycles is a significant factor that can be related to orderly petal-like structure with the accessible active

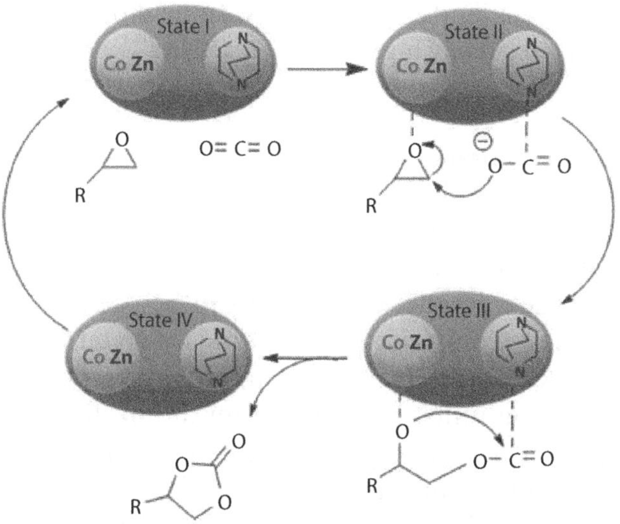

Scheme 10.1 Suggested mechanism of carbon dioxide cycloaddition with epoxide by bimetal mixed MOFs.

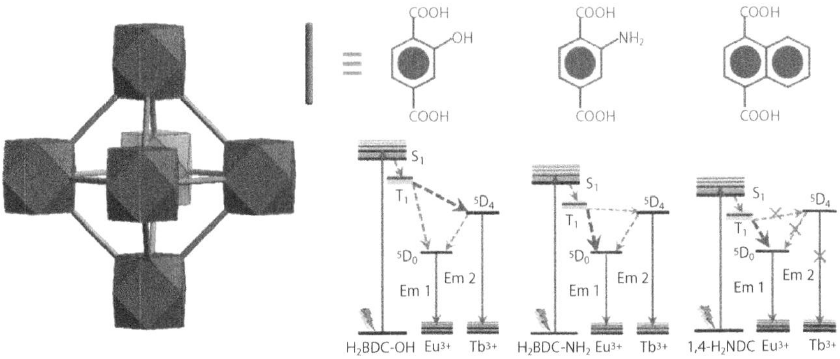

Figure 10.4 Isostructural Eu/Tb mixed MOFs with different triplet energy levels [51].

sites. So, structure and design are effective factors in the creation of new materials.

Recently, isomorphic Eu/Tb mixed MOFs have been reported by Zhao and co-workers. Three different thermo-responsive fluorescent thermometers were prepared by regulating triplet energy level of organic linkers in isostructural Eu/Tb mixed MOFs. Among them, a temperature-dependent fluorescence behavior was seen in LnBDC-NH_2. This observation is also quite rare and unusual. The results showed that $Eu_{0.01}Tb_{0.99}NDC$ is useful in the physiological range. Also, these kinds of materials with heterogeneous strcuture demonstrate a relatively high sensitivity [51].

10.2 Conclusion

The complexity in materials can appear to different forms with advances in their capabilities. In MOF-based materials, mixed MOFs are another group that can be considered as complex structures. Complexity in this kind of material can be due to the presence of different agents in structure such as multiple kinds of building units. Deviation from uniformity and symmetry is important factor in producing more complex MOF structures. So, design of materials and modification of their ability have been attracted. However, there are not many reports about mixed MOFs, despite they make incredible jumps in the useful applications. So, in this chapter, we implied to their synthesis and properties along with related examples.

References

1. Fenton, J.L., Steimle, B.C., Schaak, R.E., Tunable intraparticle frameworks for creating complex heterostructured nanoparticle libraries. *Science*, *360*, 6388, 513–517, 2018.
2. Yu, L., He, C., Zheng, Q., Feng, L., Xiong, L., Xiao, Y., Dual Eu-MOFs based logic device and ratiometric fluorescence paper microchip for visual H_2O_2 assay. *J. Mater. Chem. C*, *8*, 10, 3562–3570, 2020.
3. Schlüsener, C., Jordan, D.N., Xhinovci, M., Ntep, T.J.M.M., Schmitz, A., Giesen, B., Janiak, C., Probing the limits of linker substitution in aluminum MOFs through water vapor sorption studies: mixed-MOFs instead of mixed-linker CAU-23 and MIL-160 materials. *Dalton Trans.*, *49*, 22, 7373–7383, 2020.
4. Xu, C., Bao, M., Ren, J., Zhang, Z., NH 2-MIL-88B (Fe α In 1− α) mixed-MOFs designed for enhancing photocatalytic Cr(vi) reduction and tetracycline elimination. *RSC Adv.*, *10*, 64, 39080–39086, 2020.
5. Guan, H., Wang, N., Feng, X., Bian, S., Liu, Y., Ma, M., Li, W., Chen, Y., S element-doped synergistically well-mixed MOFs as highly efficient oxygen precipitation electrocatalyst. *Int. J. Hydrogen Energy*, *45*, 46, 24333–24340, 2020.
6. Zacher, D., Shekhah, O., Wöll, C., Fischer, R.A., Thin films of metal–organic frameworks. *Chem. Soc. Rev.*, *38*, 5, 1418–1429, 2009.
7. Yoo, Y. and Jeong, H.-K., Design, Heteroepitaxial Growth of Isoreticular Metal- Organic Frameworks and Their Hybrid Films. *Cryst. Growth Des.*, *10*, 3, 1283–1288, 2010.
8. Jiang, S., Du, Y., Marcello, M., Corcoran Jr., E.W., Calabro, D.C., Chong, S.Y., Chen, L., Clowes, R., Hasell, T., Cooper, A.I., Inside Cover: Core–Shell Crystals of Porous Organic Cages. *Angew. Chem. Int. Ed.*, *57*, 35, 11082–11082, 2018.
9. Zhang, Z., Sun, N., Wei, W., Sun, Y., Facilely controlled synthesis of a core-shell structured MOF composite and its derived N-doped hierarchical porous carbon for CO 2 adsorption. *RSC Adv.*, *8*, 38, 21460–21471, 2018.
10. Boissonnault, J.A., Wong-Foy, A.G., Matzger, A.J., Core–Shell Structures Arise Naturally During Ligand Exchange in Metal–Organic Frameworks. *J. Am. Chem. Soc.*, *139*, 42, 14841–14844, 2017.

11. Wannapaiboon, S., Tu, M., Sumida, K., Khaletskaya, K., Furukawa, S., Kitagawa, S., Fischer, R.A., Hierarchical structuring of metal-organic framework thin-films on quartz crystal microbalance (QCM) substrates for selective adsorption applications. *J. Mater. Chem. A*, *3*, 46, 23385–23394, 2015.
12. Fu, Y.Y., Yang, C.X., Yan, X.P., Fabrication of ZIF-8@SiO2 core–shell microspheres as the stationary phase for high-performance liquid chromatography. *Chem.-Eur. J.*, *19*, 40, 13484–13491, 2013.
13. Hu, P., Morabito, J.V., Tsung, C.-K., Core–shell catalysts of metal nanoparticle core and metal–organic framework shell. *ACS Catal.*, *4*, 12, 4409–4419, 2014.
14. Jahan, M., Bao, Q., Loh, K.P., Electrocatalytically active graphene-porphyrin MOF composite for oxygen reduction reaction. *J. Am. Chem. Soc.*, *134*, 15, 6707–6713, 2012.
15. Lu, G., Li, S., Guo, Z., Farha, O.K., Hauser, B.G., Qi, X., Wang, Y., Wang, X., Han, S., Liu, X., Imparting functionality to a metal–organic framework material by controlled nanoparticle encapsulation. *Nat. Chem.*, *4*, 4, 310, 2012.
16. Morabito, J.V., Chou, L.-Y., Li, Z., Manna, C.M., Petroff, C.A., Kyada, R.J., Palomba, J.M., Byers, J.A., Tsung, C.-K., Molecular encapsulation beyond the aperture size limit through dissociative linker exchange in metal–organic framework crystals. *J. Am. Chem. Soc.*, *136*, 36, 12540–12543, 2014.
17. Zhu, Q.-L. and Xu, Q., Metal–organic framework composites. *Chem. Soc. Rev.*, *43*, 16, 5468–5512, 2014.
18. Choi, S., Kim, T., Ji, H., Lee, H.J., Oh, M., Isotropic and Anisotropic Growth of Metal–Organic Framework (MOF) on MOF: Logical Inference on MOF Structure Based on Growth Behavior and Morphological Feature. *J. Am. Chem. Soc.*, *138*, 43, 14434–14440, 2016.
19. Furukawa, S., Hirai, K., Nakagawa, K., Takashima, Y., Matsuda, R., Tsuruoka, T., Kondo, M., Haruki, R., Tanaka, D., Sakamoto, H., Heterogeneously hybridized porous coordination polymer crystals: fabrication of heterometallic core–shell single crystals with an in-plane rotational epitaxial relationship. *Angew. Chem. Int. Ed.*, *48*, 10, 1766–1770, 2009.
20. Koh, K., Wong-Foy, A.G., Matzger, A.J., MOF@MOF: microporous core–shell architectures. *Chem. Commun.*, 41, 6162–6164, 2009.
21. Li, T., Sullivan, J.E., Rosi, N.L., Design and preparation of a core–shell metal–organic framework for selective CO_2 capture. *J. Am. Chem. Soc.*, *135*, 27, 9984–9987, 2013.

22. Hirai, K., Furukawa, S., Kondo, M., Meilikhov, M., Sakata, Y., Sakata, O., Kitagawa, S., Targeted functionalisation of a hierarchically-structured porous coordination polymer crystal enhances its entire function. *Chem. Commun.*, *48*, 52, 6472–6474, 2012.
23. Wang, L., Yang, W., Li, Y., Xie, Z., Zhu, W., Sun, Z.-M., Dynamically controlled one-pot synthesis of heterogeneous core–shell MOF single crystals using guest molecules. *Chem. Commun.*, *50*, 79, 11653–11656, 2014.
24. Fukushima, T., Horike, S., Kobayashi, H., Tsujimoto, M., Isoda, S., Foo, M.L., Kubota, Y., Takata, M., Kitagawa, S., Modular design of domain assembly in porous coordination polymer crystals via reactivity-directed crystallization process. *J. Am. Chem. Soc.*, *134*, 32, 13341–13347, 2012.
25. Yang, J., Zhang, F., Lu, H., Hong, X., Jiang, H., Wu, Y., Li, Y., Hollow Zn/Co ZIF Particles Derived from Core–Shell ZIF-67@ ZIF-8 as Selective Catalyst for the Semi-Hydrogenation of Acetylene. *Angew. Chem. Int. Ed.*, *54*, 37, 10889–10893, 2015.
26. Liu, J. and Wöll, C., Surface-supported metal–organic framework thin films: fabrication methods, applications, and challenges. *Chem. Soc. Rev.*, *46*, 19, 5730–5770, 2017.
27. Li, M. and Dincă, M., Selective formation of biphasic thin films of metal–organic frameworks by potential-controlled cathodic electrodeposition. *Chem. Sci.*, *5*, 1, 107–111, 2014.
28. Fuoco, A., Khdhayyer, M., Attfield, M., Esposito, E., Jansen, J., Budd, P., Synthesis and transport properties of novel MOF/PIM-1/MOF sandwich membranes for gas separation. *Membranes*, *7*, 1, 7, 2017.
29. Allendorf, M., Bauer, C., Bhakta, R., Houk, R.J.T., Luminescent metal–organic frameworks. *Chem. Soc. Rev.*, *38*, 5, 1330–1352, 2009.
30. Evans, O.R. and Lin, W., Crystal engineering of NLO materials based on metal–organic coordination networks. *Acc. Chem. Res.*, *35*, 7, 511–522, 2002.
31. Czaja, A.U., Trukhan, N., Müller, U., Industrial applications of metal–organic frameworks. *Chem. Soc. Rev.*, *38*, 5, 1284–1293, 2009.
32. Guo, H., Zhu, G., Hewitt, I.J., Qiu, S., "Twin copper source" growth of metal- organic framework membrane: $Cu_3(BTC)_2$ with high permeability and selectivity for recycling H_2. *J. Am. Chem. Soc.*, *131*, 5, 1646–1647, 2009.
33. Bux, H., Liang, F., Li, Y., Cravillon, J., Wiebcke, M., Caro, J., Zeolitic imidazolate framework membrane with molecular sieving properties

by microwave-assisted solvothermal synthesis. *J. Am. Chem. Soc.*, *131*, 44, 16000–16001, 2009.
34. Shekhah, O., Liu, J., Fischer, R., Wöll, C., MOF thin films: existing and future applications. *Chem. Soc. Rev.*, *40*, 2, 1081–1106, 2011.
35. Lu, G. and Hupp, J.T., Metal–organic frameworks as sensors: a ZIF-8 based Fabry–Pérot device as a selective sensor for chemical vapors and gases. *J. Am. Chem. Soc.*, *132*, 23, 7832–7833, 2010.
36. Shekhah, O. and Eddaoudi, M., The liquid phase epitaxy method for the construction of oriented ZIF-8 thin films with controlled growth on functionalized surfaces. *Chem. Commun.*, *49*, 86, 10079–10081, 2013.
37. Furukawa, S., Hirai, K., Takashima, Y., Nakagawa, K., Kondo, M., Tsuruoka, T., Sakata, O., Kitagawa, S., A block PCP crystal: anisotropic hybridization of porous coordination polymers by face-selective epitaxial growth. *Chem. Commun.*, 34, 5097–5099, 2009.
38. Liu, B., Shekhah, O., Arslan, H.K., Liu, J., Wöll, C., Fischer, R.A., Enantiopure metal–organic framework thin films: oriented SURMOF growth and enantioselective adsorption. *Angew. Chem. Int. Ed.*, *51*, 3, 807–810, 2012.
39. Liu, B., Tu, M., Fischer, R.A., Metal–organic framework thin films: crystallite orientation dependent adsorption. *Angew. Chem. Int. Ed.*, *52*, 12, 3402–3405, 2013.
40. Wu, Y., Song, X., Xu, S., Chen, Y., Oderinde, O., Gao, L., Wei, R., Xiao, G., Chemical fixation of CO_2 into cyclic carbonates catalyzed by bimetal mixed MOFs: the role of the interaction between Co and Zn. *Dalton Trans.*, *49*, 2, 312–321, 2020.
41. Raja, D.S., Huang, C.L., Chen, Y.A., Choi, Y., Lu, S.Y., Composition-balanced trimetallic MOFs as ultra-efficient electrocatalysts for oxygen evolution reaction at high current densities. *Appl. Catal. B: Environ.*, *279*, 119375, 2020.
42. Yuan, J.T., Hou, J.J., Liu, X.L., Feng, Y.R., Zhang, X.M., Optimized trimetallic benzotriazole-5-carboxylate MOFs with coordinately unsaturated active sites as an efficient electrocatalyst for the oxygen evolution reaction. *Dalton Trans.*, *49*, 3, 750–756, 2020.
43. Huo, J., Wang, Y., Yan, L., Xue, Y., Li, S., Hu, M., Jiang, Y., Zhai, Q.G., *In situ* semi-transformation from heterometallic MOFs to Fe–Ni LDH/MOF hierarchical architectures for boosted oxygen evolution reaction. *Nanoscale*, *12*, 27, 14514–14523, 2020.
44. Zhang, M., Lin, Q., Wu, W., Ye, Y., Yao, Z., Ma, X., Xiang, S., Zhang, Z., Isostructural MOFs with Higher Proton Conductivity for Improved

Oxygen Evolution Reaction Performance. *ACS Appl. Mater. Interfaces, 12*, 14, 16367–16375, 2020.

45. Goswami, A., Ghosh, D., Chernyshev, V.V., Dey, A., Pradhan, D., Biradha, K., 2D MOFs with Ni(II), Cu(II), and Co(II) as Efficient Oxygen Evolution Electrocatalysts: Rationalization of Catalytic Performance vs Structure of the MOFs and Potential of the Redox Couples. *ACS Appl. Mater. Interfaces, 12*, 30, 33679–33689, 2020.
46. Xue, J.Y., Li, C., Li, F.L., Gu, H.W., Braunstein, P., Lang, J.P., Recent advances in pristine tri-metallic metal–organic frameworks toward the oxygen evolution reaction. *Nanoscale, 12*, 8, 4816–4825, 2020.
47. Zou, Z., Wang, J., Pan, H., Li, J., Guo, K., Zhao, Y., Xu, C., Enhanced oxygen evolution reaction of defective CoP/MOF-integrated electrocatalyst by partial phosphating. *J. Mater. Chem. A, 8*, 28, 14099–14105, 2020.
48. Zhao, S., Tan, C., He, C.T., An, P., Xie, F., Jiang, S., Zhu, Y., Wu, K.H., Zhang, B., Li, H., Zhang, J., Structural transformation of highly active metal–organic framework electrocatalysts during the oxygen evolution reaction. *Nat. Energy*, 5, 881–890, 2020.
49. Ji, Q., Kong, Y., Wang, C., Tan, H., Duan, H., Hu, W., Li, G., Lu, Y., Li, N., Wang, Y., Tian, J., Lattice Strain Induced by Linker Scission in Metal–Organic Framework Nanosheets for Oxygen Evolution Reaction. *ACS Catal., 10*, 10, 5691–5697, 2020.
50. Wang, H.F., Chen, L., Pang, H., Kaskel, S., Xu, Q., MOF-derived electrocatalysts for oxygen reduction, oxygen evolution and hydrogen evolution reactions. *Chem. Soc. Rev., 49*, 5, 1414–1448, 2020.
51. Xia, T., Shao, Z., Yan, X., Liu, M., Yu, L., Wan, Y., Chang, D., Zhang, J., Zhao, D., Tailoring the triplet level of isomorphic Eu/Tb mixed MOFs for sensitive temperature sensing. *Chem. Commun., 57*, 25, 3143–3146, 2021

Index

Achiral, 19, 106, 107, 108, 109, 115, 116, 120, 127, 128, 130, 131, 132, 136, 137, 144
Alkynylzinc addition, 114
Asymmetric, 18, 59, 65, 105, 106, 107, 110, 112, 114, 115, 116, 120, 121, 127, 129, 130, 132, 135, 136, 137, 139, 140, 141, 142, 144

Brønsted (acid, acidity, base, basicity), 137, 139, 142, 156
Building block(s), 2, 6, 14, 15, 16, 17, 18, 21, 27, 62, 73

Camphorate, 106, 115, 116
Carboxylate(s), 1, 2, 37, 43, 57, 59, 157, 173
Chiral, 14, 16, 17, 18, 19, 21, 43, 64, 74, 78, 105, 106, 107, 108, 109, 110, 111, 112, 114, 115, 116, 117, 118, 120, 121, 127, 128, 129, 130, 131, 132, 133, 134, 135, 136, 137, 138, 139, 141, 142, 143, 144, 189
CO_2 sorption, 43
Complexity, 7, 8, 13, 14, 15, 16, 17, 18, 19, 20, 21, 22, 27, 28, 35, 36, 43, 51, 57, 58, 59, 60, 61, 62, 63, 64, 71, 72, 73, 81, 92, 105, 109, 111, 115, 116, 117, 120, 127, 130, 131, 136, 137, 139, 149, 154, 171, 172, 173, 179, 185, 186, 187, 188, 189, 192
Composite building units (CBUs), 35
Coordination polymers, 1, 15, 18, 59, 60, 77, 78, 115, 116
Core-shell, 78, 79, 80, 185, 186, 187
Crystalline MOF(s), 57, 66, 109, 150, 154, 155
Cycloaddition, 90, 190, 191

Defect (s), 14, 16, 17, 19, 20, 21, 37, 38, 40, 42, 66, 106, 149, 150, 151, 152, 153, 154, 155, 156, 157, 158, 159, 160, 161, 162, 174
Diethylzinc addition(s), 113, 114,
Direct synthesis, 6, 16, 50, 59, 60, 73, 74, 106
Ditopic linker(s), 28, 33, 34, 43, 60
Diversity, 6, 13, 18, 59, 61, 63, 71, 131, 150, 154

Enantiomer(s), 106, 127, 129, 132
Epitaxial growth, 78, 79, 186, 187

Functional group(s), 2, 3, 17, 20, 28, 33, 60, 86, 110, 111, 131, 132, 136, 137, 139, 143, 144, 160, 186, 187, 189
Functionalization, 1, 20, 60, 131, 157, 189

Heterochiral, 17, 108, 109
Heterogeneity, 7, 8, 13, 14, 15, 16, 17, 19, 20, 21, 27, 28, 33, 38, 43, 71, 81, 86, 87, 120, 136, 139, 149, 150, 154, 155, 161, 162, 171, 172, 173, 175, 177, 179, 185, 187, 189

Heterogeneous, 8, 13, 14, 15, 16, 17, 21, 34, 41, 43, 46, 49, 65, 77, 81, 82, 114, 115, 129, 137, 143, 149, 154, 160, 161, 162, 171, 172, 173, 174, 175, 176, 177, 179, 192
Hexagonal channels, 37, 64
Homochiral, 105, 106, 107, 108, 109, 116, 128
Homogeneity, 20, 27, 43, 143
Hydrothermal, 6

Immobilization, 72, 79, 136, 143, 188, 189
Incorporation, 3, 14, 20, 34, 37, 40, 43, 107, 118, 120, 136, 143, 160
Indirect synthesis, 6, 131
Inherent chirality, 16, 105, 106, 107
Inherent defect(s), 16, 149, 150, 153, 154
Interpenetration(s), 63, 64, 83, 119, 153, 158, 162

Linker exchange, 37, 38, 39, 154
Linker labilization, 37, 38, 39, 41
Liquid-phase epitaxy, 79, 189

Macropore(s), 171, 173, 175, 176, 177, 179
Mesopore(s), 37, 40, 41, 66, 156, 158, 171, 173, 176, 177, 179
Metal exchange, 71, 72, 78, 79, 80, 83
Metalloligand(s), 73, 74, 77, 78, 80, 116, 118
Metallosen ligand(s), 34
Micropore(s), 40, 41, 66, 171, 173, 174, 175, 177
Missing linker(s), 150, 153, 154, 155, 174
Missing node(s), 153, 174
Mixed component(s), 15, 27, 78
Mixed ligand(s), 16, 18, 19, 27, 28, 29, 30, 31, 32, 33, 34, 37, 41, 43, 45, 46, 50, 51, 59, 60, 120, 155, 187
Mixed metal(s), 14, 16, 18, 19, 20, 21, 71, 81, 88, 89, 90, 91, 187
Mixed valence metals, 20
Mixed-MOFs, 15
MOF-on-MOF, 78, 79, 185, 186, 189
Multi-heterotopic ligands (linkers), 16, 17, 28, 57, 63, 64, 66
Multiple SBUs, 14, 16, 21, 58, 71, 81, 86, 91
Multi-topic linker, 62
Multivariate MOF, 59

Non-porous, 13, 36

Olefin(s), 41, 42, 89, 118, 129, 130, 139, 142
Organic linker(s), 1, 2, 3, 4, 5, 7, 13, 15, 18, 19, 20, 21, 28, 34, 36, 38, 39, 40, 43, 57, 59, 61, 63, 64, 66, 71, 72, 73, 81, 86, 87, 91, 107, 109, 115, 118, 120, 131, 132, 134, 153, 155, 156, 160, 173, 192

Polytopic linker(s), 16, 17, 57, 58, 59, 60
Post-synthesis modification, 3, 134, 138

Space group, 45, 64, 82, 87, 107, 121, 132
Spontaneous resolution, 106, 107, 108
Symmetry, 17, 18, 57, 58, 59, 61, 62, 63, 64, 66, 120, 192

Template(s), 16, 106, 127, 128, 129, 130, 131, 144, 155, 172, 174
Topology, 1, 17, 21, 45, 49, 64, 115, 116, 117, 120, 155, 173, 174
Transmetalation, 71, 78, 80
Triangular channels, 37
Tritopic linker(s), 58, 85
Tunability, 1, 3, 17, 27, 37, 45, 63, 186

Also of Interest

Check out the previously published books by the same author

Metal-Organic Frameworks with Heterogeneous Structures
By Ali Morsali and Kayhaneh Berijani
Published 2021. ISBN 978-1-119-79204-8

Functionalized Metal-Organic Frameworks
By Ali Morsali and Sayed Ali Akbar Razavi
Published 2020. ISBN 978-1-119-64043-1

Pillared Metal-Organic Frameworks: Properties and Applications
By Lida Hashemi and Ali Morsali
Published 2019. ISBN 978-1-119-46024-4

Main Group Metal Coordination Polymers: Structures and Nanostructures
By Ali Morsali and Lida Hashemi
Published 2017. ISBN 978-1-119-37023-9

www.scrivenerpublishing.com

Printed and bound by CPI Group (UK) Ltd, Croydon, CR0 4YY
17/09/2026
14964247-0001